HEROES

ALSO BY PAUL JOHNSON

Creators

George Washington

Art: A New History

A History of the American People

The Quest for God

The Birth of the Modern

Intellectuals

A History of the English People

A History of the Jews

Modern Times

A History of Christianity

HE

ROES

From Alexander the Great and Julius Caesar to Churchill and de Gaulle

PAUL JOHNSON

An Imprint of HarperCollins*Publishers*
www.harpercollins.com

HarperCollins books may be purchased for educational, business, or sales promotional use. For information, please write: Special Markets Department, HarperCollins Publishers, 10 East 53rd Street, New York, NY 10022.

FIRST EDITION

Designed by Emily Cavett Taff

Library of Congress Cataloging-in-Publication Data is available upon request.

ISBN: 978-0-06-114316-8
ISBN-10: 0-06-114316-2

07 08 09 10 11 OV/RRD 10 9 8 7 6 5 4 3 2 1

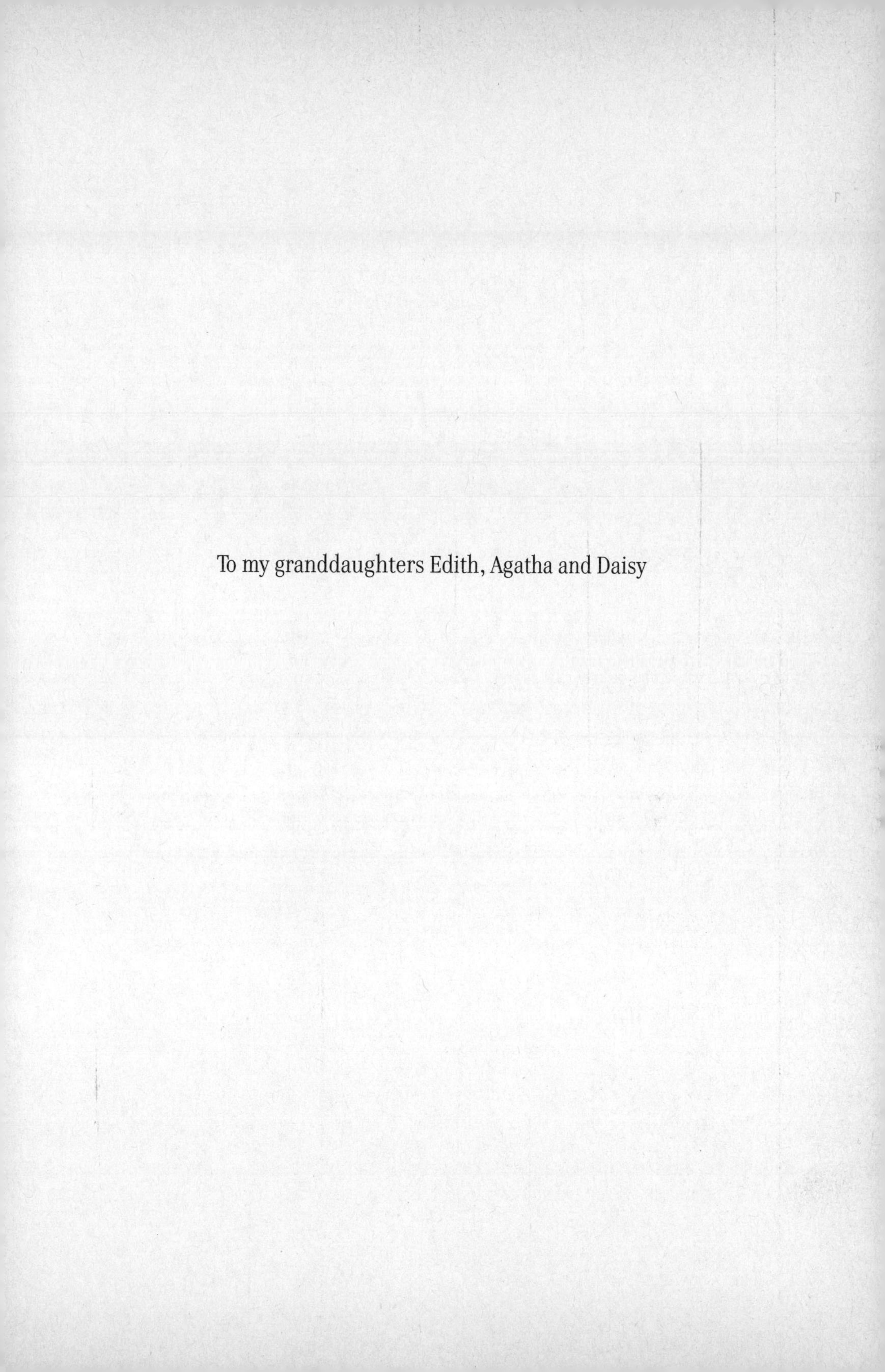

To my granddaughters Edith, Agatha and Daisy

CONTENTS

INTRODUCTION

WHAT IS A HERO?

In the Western Desert of Egypt, some twenty miles from the nearest road, you come across solitary tombs of stone, much weathered by the wind. Some are buried in the sand. They may be 3,000 or even 4,000 years old and testify to the veneration once felt for men of outstanding virtue or generosity or heroism by their younger contemporaries, who built the tombs to mark their respect and perpetuate the memory of the dead. The names have long since been obliterated by time and weather, but a certain sanctity hovers around the spot still. So too, in the Alps and the Tyrol, in the Pyrenees and the Carpathians, little shrines by the wayside, and rustic ornamental fountainheads over springs, commemorate the lives of local men or women who once struck their neighbors as remarkable. Occasionally the name survives. Usually time has imposed anonymity. But the spirit of virtue hovers over the illustrious dead. The names do not matter. It is the principle of honoring the good or the brave which strikes us. Dr. Johnson records in his *A Journey to the Western Islands* his visit to Iona, and in particular what he believed to be the cemetery of the ancient Scottish kings and other famous men. The place filled him with awe and he commented: "By whom the subterraneous vault are peopled is now unknown. The graves are very numerous and some of them undoubtedly contain the remains of men who did not expect to be so soon forgotten."

I have visited Iona, and meditated like Dr. Johnson on these sepulchres of unknown notabilities. Such places stir the imagination and make me think more kindly of the human race. I have seen such shrines in Greece, some going back to before the Greeks came. Homer called those thus honored Υρω-ες, heroes, defined as "a name given to men of superhuman strength, courage or ability, favoured by the gods; at a later time regarded as intermediate between gods and men, and immortal." Graveyards of distant kings are always impressive, and I relish seeing them: the great Fourth Dynasty pharaohs at Giza, for example, or the tombs of the Holy Roman emperors in the Palermo cathedral, especially the vast but simple block of black marble which covers the last resting place of the most formidable of them all, Henry VI; or the tombs of the Angevins in the abbey church of Fontevrault, between Chinon and Poitiers. There rest our great law-giving King Henry II, his wife Eleanor of Aquitaine, his son Richard the Lionheart, and Isabella, second wife of Henry's bad son, King John. The remains of France's kings were once carefully preserved in the royal abbey of Saint-Denis, their hearts being kept separately in reliquaries. But all was desecrated by the sansculottes of the revolution, precious items being sold off for cash: thus the shriveled heart of Louis XIV, the Sun King, ended up at Stanton Harcourt, where it was sacrelegiously eaten by a Cambridge professor. The sanctuary of the English kings, Westminster Abbey, has fared better, and still houses intact the remains of its founder, Edward the Confessor, and many of his successors.

Where kings lie, other famous people will be buried, and the abbey is crowded with the tombs of admirals, generals and other celebrities. From the time of Chaucer it became an intermittent custom to bury poets in one corner, and the practice was extended to other writers, who are sometimes buried, more often commemorated, among the poets. Such heroes do not necessarily have to have been believers in the Christian faith of the abbey—even Charles Darwin, patron saint of the atheists, is there in spirit. Such collective cenotaphs of heroes hover uneasily between Christianity and paganism. In Paris, the ancient church of Sainte-Geneviève, patron saint of the city, founded by Clovis and rebuilt by Louis XV, was transformed by the revolutionaries in 1791 into a secular vault of heroes, and renamed the Panthéon. Reconsecrated at the restoration in 1815, it oscillated repeatedly in the nineteenth century between a church and a hero house, deconsecrated in 1885 to receive the body of Victor Hugo, who, after much vacillation on his deathbed, finally insisted that he died outside the church. Today it is a musty junk shop of the dead, not particularly edifying to those of any faith or none. In Germany, the pagan-minded nationalists had the neoclassical architect Leo von Klenze build them a Walhalla (or Valhalla) for heroes in the shape of a Greek temple (1830–1842). It is near Donanstanf and overlooks the Rhine, but it did not capture the Teutonic imagination. In Spain, after the civil war, the victor General Franco built a valley of the Fallen for heroes. But this is tainted with politics. The United States has a two-tier heroic commemoration: the great presidents, such as Washington and Lincoln, are carved out of the rocks at Mount Rushmore, while more everyday heroes repose in Arlington National Cemetery, outside the federal capital.

We should not take it for granted that the original heroes were all men, any more than we should assume that the primitive races worshipped only gods. Goddesses make their appearance in the archeological evidence from the very earliest times, and we can be sure that heroines followed swiftly in the steps of heroes. Humanity invented gods as the originators of natural events they could not understand, and feared, and gods were therefore terrible personages. What ordinary mortals needed to identify with were creatures, recognizably human but of great capacity and accomplishment, who stood halfway between the deities and the rest. These demigods were heroes, and they had to include heroines, like Pallas and Medea, for pure goddesses were too frightening to be domesticated and reduced to human scale. Once heroines come into existence, even if, like Medea, they sometimes take the form of witches, the concept of the hero ceases to be the exclusive preserve of military men or those who rejoice in superhuman physical strength. For the purpose of this collection of biographical essays, I have taken the concept of the heroic individual in its widest possible sense, even if I have included a number famed for military exploits, such as Alexander and Caesar among the males, and Boudica and Joan of Arc among the females. The fact is, anyone is a hero who has been widely, persistently over long periods, and enthusiastically regarded as heroic by a reasonable person, or even an unreasonable one. I have put into this collection one or two heroes and heroines of my own, believing that an element of idiosyncrasy is a legitimate part of hero worship. Indeed it is only by asking ourselves how we, personally, judge heroism that we begin to get to the essence of the matter. It was Madame Cornuel (1605–1694) who first noted: "*Il n'y a point de héros pour son valet de chambre.*" The point was deliberately contradicted by Lord Byron in "Beppo":

> *. . . he was a perfect cavaliero*
> *And to his very valet seemed a hero.*

Byron himself was, certainly, no hero to his valet Fletcher, who had many sensible things to say about his master's weaknesses, though devoted to him nonetheless. But some valets are heroes to their masters, at any rate fictional ones. Thus Crichton, the butler, emerges as the hero of J. M. Barrie's play *The Admirable Crichton* when all are put to the test, and Jeeves is from start to finish the hero of P. G. Wodehouse's Bertie Wooster. As Fletcher noted, Byron was quite capable of deliberately performing a heroic act. Thus we find him writing to Henry Drury from an address given as "*Salsette* Frigate, in the Dardanelles, off Abydos, 3 May 1810," describing his imitation of Leander's classic feat:

> This morning I *swam* from *Sestos* to *Abydos*—the immediate distance is not above a mile but the current renders it

> hazardous, so much so that I doubt whether Leander's conjugal powers must not have been exhausted in his passage to Paradise—I attempted it a week ago and failed owing to the north wind and the wonderful rapidity of the tide. . . . But this morning, being calmer, I succeeded and crossed "the Broad Hellespont" in an hour and ten minutes.

It is an interesting point, and Byron provides a good example, that a hero does not stand still in popular estimation. Both in his lifetime and since Byron moved in and out of heroic status, though in some countries, especially Greece and Italy, his heroism rating has seemed more secure than in others, especially Britain. Another example is Rupert Brooke, a superhero in the First World War who is disregarded as ridiculous now. It has always been so. Herod, in his time, was certainly regarded as a hero by the countless thousands of Jews and near-Jews who benefited from his munificence. The richest man in the Roman Empire and perhaps the outstanding philanthropist of its entire history, his public works—ports, markets, stadia, theaters, lighthouses, roads, housing estates and gardens—were benefits much appreciated: hence his popular title "the Great." But his Massacre of the Innocents did for all that, and once the Christians began to mold opinion he became a monster. But equally, monsters can be transformed into heroes. Genghis Khan, reviled for nearly a millennium as the archetypal mass murderer and rapist, despoiler, arsonist and ravager, has in the last two decades, since the collapse of the Soviet empire in Central Asia, become there a state-sponsored hero, especially in Azerbaijan and in Mongolia. Statues of him have arisen, towns are named in his honor and children are taught to regard him as the father of their country.

Hero movements up and down are usually less startling but frequent and continuous. Tom Jones, in his *Whitehall Diaries,* quotes Stanley Baldwin, when prime minister, remarking:

> Contemporary judgments were illusory; look at Lincoln's case, how in his lifetime he was thought to be a clumsy

> lumbering countryman, blundering along without knowing where he was going. Since his death his significance has grown steadily. [Woodrow] Wilson, on the other hand, was for a short spell looked up to like a god, and his fame will gradually shrink. Lincoln is Wisdom, and Wilson is Knowledge.

In some cases, heroic categories have been downgraded in their entirety. "Stout" Cortez and Pizarro are today regarded as the worst kind of imperialists, and cruel too. Livingstone, an outstanding hero of the Victorians, and still venerated in my childhood, is now described as a racist, and his interlocutor Stanley, hero of one of the most famous meetings in history, as a mercenary journalist. In general, explorers and travelers enjoy fewer kudos than formerly. Amundsen and Peary, even the tragic-heroic Scott of the Antarctic and his self-sacrificing companion Captain Oates, no longer occupy the same high rank in the pantheons of schoolboys. All those concerned with the spread of empire—Clive of India, Marshal Lyautey of Morocco, Cecil Rhodes, Lawrence of Arabia—are now suspect.

What follows in this book is a small selection of heroes and heroines who still evoke wonder or admiration or respect or in some cases sympathy. Some names are obvious, others less so. As with my book *Creators,* I am trying to approach the subject of heroism not so much by definition and analysis as by example. There are some fighting heroes here but I take on board the critical tradition, lamented by Erasmus in the early sixteenth century and continued notably by Swift in the eighteenth, that admiring heroic killers has its dangers. Richardson summed up this attitude when he wrote: "The fierce fighting *Iliad* has done infinite mischief for a series of ages; since to it and its copy the *Aeneid,* is owing in great measure the savage spirit that has actuated, from the earliest ages to this time, the fighting fellows, that, worse than lions and tigers, have ravaged the earth and made it a field of blood." Dr. Johnson did not go so far as to condemn Homer and Virgil, but his book on the Highlands is a commentary on the passing of the heroic chieftain and his armed warriors, which he does not regret, and he argues

> That a man, who places honour only in successful violence, is a very troublesome and pernicious animal in time of peace; and that the martial character cannot prevail in a whole people, but by the diminution of all other virtues.

The Highland personality he most enjoyed meeting, and clearly admired above all others, was Flora Macdonald, "a name that will be mentioned in history, and if courage and fidelity be virtues, mentioned with honour." He found her "a woman of middle stature, soft features, gentle manner and elegant presence."

I too have accorded women a high place in my selection, sometimes but not necessarily on account of their courage and fidelity. But my aim throughout has not been to give particular emphasis to sex or occupation but to stress that heroic behavior is to be found in every age and in all kinds of places. The chief criterion is the verdict of the public and this, being arbitrary, eccentric and often irrational (as well as changeable), gives a salty flavor to the business.

HEROES

1

GOD'S HEROES: DEBORAH, JUDITH, SAMSON AND DAVID

No people in history were more in need of heroes than the Hebrews. Cast in their role, by events, as "strangers and sojourners," they came originally in the time of Abraham, deep in the second millennium BC, from what is now southern Iraq, and entered recorded history between 1450 and 1250 BC. They were a slave people, without country and possessions, with little art or technology and no skill or record in warfare. They were subjects of the Egyptians, the greatest power of the Bronze Age, and woefully oppressed. They were not numerous either. It is one of the miracles of the human story that this tiny people, instead of disappearing into oblivion through the yawning cracks of history, as did thousands of other tribal groups—and even scores of famous nations—should still be in self-conscious existence today, an important piece on the great world chessboard, recognizably the same entity as nearly four thousand years ago.

Yet history has no miracles: only causes and consequences. And the reasons the Hebrews survived are that they had a god, a sole god, whom they worshipped with unique intensity and exclusiveness; and they had their own language, first oral then written, in which they recorded his favor and protection. They were weak in the physics of survival, strong in the metaphysics. They were first henotheists, with their sole god, Yahweh, as the divine ruler of their tribal confederation (what the Greeks called an amphictyony); then, during their Egyptian sojourn, they elevated their religious system into monotheism, Yahweh becoming sole god of the universe and all its peoples. This belief, which made them unique in antiquity, emerged under the first of their heroes, Moses, who took them out of Egyptian slavery and into independent nationhood. He gave them, in writing, their first code of divine laws, and led them, through forty years of tribulations and testing, to the edge of the land, "flowing with milk and honey," which Yahweh had promised them. Moses made full use too of their second great gift, their language. Hebrew was not only sinewy, expressive and resourceful, but peculiarly adapted to the recording of history. It was Moses, according to rabbinical tradition, who set down the first five books of the Hebrew national epic, the Pentateuch, the opening section of what became known as the Bible, to Christians the Old Testament. On this foundation the Bible accumulated over the generations, as the canonical record of Hebrew history, in time, "the most famous book in the history of the world," "*the* book," as its name implies. Moses and his doings continued to dominate it. He is mentioned 767 times in the Old Testament and 79 times even in the New Testament, the Christian title deeds. No other hero of antiquity, at any other period or from any other region, has this degree of heroic celebrity.

Moses having created, as it were, the matrix of heroism, Hebrew records arranged the continuing story of the people and heroic figures. The Book of Joshua, conservatively dated from between 1375 BC and 1045 BC, tells how the Promised Land was conquered and settled. According to the Talmud, the Jewish record of oral teaching compiled from 400 BC to AD 500, Joshua, the Hebrew general, as well as being the hero of the book, was also its author, except for a coda recording his burial. The next biblical book, Judges, is the key work in Hebrew heroism, however. It was written late in the first millennium BC, after Joshua, and supposedly, according to the Talmud, by the Prophet Samuel. By rights it should be called the Book of Heroes, for most of its celebrities were not judges but fighters, who enabled the Hebrews, or Israelites as they began to be called, to survive as a recognizable, independent people during extremely difficult times.

The last centuries of the second millennium BC, the chronological junction between the Bronze Age and the new Iron Age, witnessed one of the greatest convulsions of antiquity, involving the whole of the eastern Mediterranean and its shores and much of western Asia—huge conquests and tribal movements, invasions and dispossessions, massacres and genocides and great minglings and fusions of peoples. The destruction of Mycenaean Greece and Crete, the events reflected in Homer's later (eighth century BC) recounting of the Trojan War, the near-disintegration of the New Kingdom of Egypt and the movement of the Hebrews themselves into Palestine are all part of this reshuffling of the cards of history, the biggest upheaval until Alexander the Great broke up the Persian Empire and replaced it with the Hellenistic world.

The literary and archeological evidence of these massive but obscure events is fragmentary, and the picture historians have been able to build up is confused, and likely to remain so. No wonder, then, that the Book of Judges, recording part of the turmoil in a small area of the scene of convulsion, from a Hebrew viewpoint, is a confused and confusing document also. It claims to record the deeds of a dozen "judges," or tribal heroes, called Othniel, Ehud, Shamgar, Deborah (and her general Barak), Gideon, Tola, Jain, Jephthah, Ibzan, Elon, Abdon and Samson. But, though they are presented in chronological order, it may be that some were contemporaries. Their importance varied greatly. Several were obscure leaders of a single tribe. Others were national figures. The enemy too varied. At the beginning of the book, the enemy seems to be the original inhabitants of the Promised Land, the Canaanites. But no scholar has been able to settle the origins of this term, and it may be that it signified no more than the collective expression for a mass of small kingdoms and amphictyonies, living in Syria and Palestine: the Jebusites and Amorites, Girgashites, Hivites, Arkites, Sinites, Arradites, Zemarites, Hamathites and others. The term may mean "lowly" or "low born," and be abusive. One of the earliest references to Canaanites occurs in a clay tablet from the administrative records of the city of Mari (fifteenth century BC), which refers to "thieves and Canaanites." The Hebrews regarded them with horror, for the Canaanite group of peoples had a superior artistic culture, and superficially attractive religious cults, and Hebrews were tempted to assimilate with them and intermarry. It was the Hebrew religious instinct to maintain a strict code of racial apartheid, which made dangerous contacts as difficult as possible. The best way to uphold this doctrine was to pursue continuous warfare, in pursuit of land, booty and slaves. The Judges personified this policy.

However, the various Canaanite petty kingdoms, even including the more technically advanced ones which eventually coalesced into the people we know as the Phoenicians (producers of the alphabet we still use today), were not the only or even the principal enemy of the Hebrews. In the Old Testament, "enemies" are mentioned 919 times. Of these, 423 mentions, or 46 percent, refer to the Philistines. These formed one of the great, magisterial lost peoples of antiquity, whose name alone survives in the term Palestine, still used for the Promised Land of the Hebrews. This word was Egyptian, "Pelest," or as the Egyptians put it, having no written vowels, "Plst." The Egyptians knew them as the "Sea Peoples," invaders who arrived from the north in ships. They came close to destroying the Pharaonic kingdom of Egypt and they evoked well-grounded terror wherever they penetrated. They were tremendous warriors by sea and land, rather like the Normans in the history of early medieval Europe. They overwhelmed the Hittites in Anatolia and destroyed the ancient and heavily fortified seaport of Ugarit in Syria.

They transformed Canaan into Palestine and took and fortified five big towns, Ashdod, Ashkelon, Gaza, Gath and Ekron. They were, as we would say, Europeans rather than Asiatics or Africans. They came from Greece and Crete. With the collapse of Aegean civilization in the thirteenth century BC they had become pirates and mercenaries, but they brought with them their iron culture and Aegean-type pottery. We know what they looked like, for the Egyptian low-relief sculptors, with fear and trembling, carved their images on the temple walls of Medinet Habu. They were tall and slender—giants to most Asians—clean-shaven and eagle-eyed. They wore paneled kilts with tassels and their chests were protected by multilayered ribbed linen corselets. Their headgear, distinctive and frightening, were upright circles of reeds or leather straps or horse-hair, mounted on a close-fitting cap. Each warrior carried a pair of spears or a long sword, or both. They had three-man iron chariots, each with a driver and two spearmen, and behind their armies, their families followed in oxcarts. Their mastery of hard-metal working made them more than a match for Bronze Age peoples in battle. And in the arts too, for their skills were considerable. The term "Philistine" as a hater of art is a misnomer: the Hebrews had nothing by comparison.

This formidable people moved into the coastal strip, slaughtering the Canaanites and pushing into the interior, confining the Hebrews to the mountains and their foothills. The Hebrews, then, faced a variety of enemies whom they needed to defeat simply to survive, let alone to occupy all the rich land Yahweh had promised them. Their resources, as noted, were metaphysical rather than primarily physical—they had tremendous religious morale springing from their monotheism and their clear code of ethics. But in one respect they tapped a physical resource which most ancient peoples denied themselves: they made full use of the brains and courage of their women. How it came about that so many great peoples, until quite recently, failed to draw upon half their human capital we shall never know. But the Hebrews did not fail—just as well, since they had so little else—and the Bible is the record of their common sense. It is a curious fact that the first written record of a joke—of laughter—occurs in the Book of Genesis, and shows Sarah, Abraham's wife, in an antimasculine posture, laughing at the solemn all-male consultation between her husband, God and his angels about her approaching pregnancy. God rebukes her for laughing, but she has the last laugh, as well as the first.

This extraordinary episode, so typical of the way in which the Bible, unlike any comparable record of antiquity, continually places women in the forefront of events, sets the tone for the future. Hebrew, or Israelite, society was patriarchal, as indeed were all societies then (and indeed now), but not exclusively so. Women were prized too for their wisdom, tenderness, passion, and at times heroic ruthlessness. This is brought out with great force in the story of Deborah, which is told in Judges chapters 4 and 5. It is told twice over, first in prose, then in verse, and the Hebrew is superb. As with all the stories in Judges the scene is set by Israelite sinfulness—that is, their relaxing of racial apartheid and their mingling with the pagans, including observing their religious and cultural rites, what the Bible calls "doing what was wrong in the eyes of the Lord."

When this happens Yahweh selects an instrument for the castigation of his people, in this case "Jabez the Canaanite king, who ruled in Hazor." The account says that Jabez had a general, Sisera, "who lived in Harosheth-of-the-Gentiles," and that he oppressed the Israelites "for twenty years" (i.e., a long time, though not a very long time, which would have been "forty years"). Sisera was a mercenary, and probably a Philistine or a commander of Philistine mercenaries, who we surmise set himself up as a king in his own right. Sisera, we are told, had "nine hundred chariots of iron" and the Israelites had no mobile armor at all. But they had Deborah, and her wisdom and power of command.

This enchanting woman provides one of the most satisfying biblical portraits. She was the wife of Lappidoth, but he was a nonentity and we hear no more about him. She had many gifts and roles. First she was a prophetess. She was by no means the only woman prophet. We hear also of Miriam (Exodus 15:20), Huldah (2 Kings 22:14) and in New Testament times Anna (Luke 2:36). But Deborah was also a judge, indeed the only one of the judges who is actually described as exercising judicial functions. "It was her custom," we are told, "to sit beneath the Palm Tree of Deborah between Ramah and Bethel in the hill country of Ephrahim, and the Israelites went up to her for justice." This arcadian scene recalls Moses as judge, and evidently when this book of the Bible was edited, over two hundred years later, the tree was still in existence, and revered, and known by her name. Her evident repute and prestige as a judge reveals that she was learned, knowing all the regulations later described, not only in the Pentateuch but in Deuteronomy and Numbers, and much case law too. People came to her because her rulings were respected and took effect. When Sisera's terrifying force of iron chariots threatened the settled land, "the Israelites cried out for help" but they turned to Deborah for advice and decisions. Her ruling was prompt. She could decide, from her wisdom, the nature of the campaign to be fought against Sisera, and the general strategy. But, being a woman (and probably an old one), it was unbecoming for her to direct detailed, tactical operations on the battlefield. For that a professional soldier was needed. "So she sent for Barak, son of Abinoam from Kedesh in Naphtali," and issued to him God's commands, she acting as prophetic spokeswoman for the Deity: "Go and recruit ten thousand men from Naphtali and Zebulun, and bring them with you to Mount Tabor. I will entice Sisera . . . to the torrent of Koshon with all his chariots and his horde, and there I will deliver them into your hands."

General Barak's willingness to obey Deborah's summons testifies to her authority, and he accepted her plan moreover. But the reply with which he qualified his submission is still more telling: "If you go with me [into battle], I will go. But if you will not go, neither will I." That was blunt: her morale-boosting presence on the battlefield was essential to victory, in his view. And he, as battle commander, needed her physical reassurance, and advice on tactics too. So it had been with Moses. She assented with a grim feminist note: "Certainly I will go with you, but this action will bring you no glory, because the Lord will leave Sisera to fall into the hands of a woman."

So Deborah went with Barak at the head of his ten thousand men. When Sisera heard of Barak's movement, he took his entire force to the bottom of Mount Tabor. That was exactly what Deborah had hoped for. Torrential rains, pouring down the slopes, had turned the plain below Mount Tabor into a quagmire. She woke the sleeping Barak: "'Up! This day the Lord gives Sisera into your hands! Already the Lord has gone out to battle before you.' By this she meant the rain." So Barak came charging down from Mount Tabor with ten thousand infantry at his back. Sisera's huge force of chariots became useless in the rapidly forming marsh, sticking in the mud. Their spearmen had to dismount, and were picked off one by one. They tried to flee, but the Israelite foot soldiers pursued them, and killed all.

Sisera too abandoned his bogged-down chariot and "fled on foot." It is always a poignant moment when the commander of a powerful and triumphant cavalry force miscalculates, sees his squadrons distintegrate and suddenly finds himself alone, without even a horse. Some hours elapsed, and many weary miles. The proud commander, now muddy, frightened and exhausted, came across a group of tents of a tribe he believed friendly to King Jabin. He approached a woman's tent, for safety, and Jael came out to meet him and said: "Come in here, my lord, come in—do not be afraid." He went in, and she covered him with a rug. It was, of course, against all etiquette for a man, especially a fugitive, to violate the sanctity of a woman's tent. And Sisera, in his distress, went on to commit two further breaches of social laws. He asked for refreshment without waiting for an invitation. He said to Jael: "Give me some water to drink—I am thirsty." So she opened a skin full of milk, gave him a drink, and covered him up again. Thus emboldened, he tried to take charge of the woman. He said to her: "Stand at the tent entrance, and if anybody comes and asks if someone is here, say No." This was too much. Jael, whose husband was Heber the Kenite (another nonentity), affronted and angered, waited till Sisera was asleep, then "took a tent-peg, picked up a mallet, crept up to him and drove the peg into his skull as he slept. His brains oozed out into the ground, his limbs twitched, and he died." In due course Barak arrived in pursuit, and Jael went out to him and said, "Come, I will show you the man you are looking for." Barak went in, found the wretched corpse, and remembered Deborah's prophecy.

This is a grim but fascinating and convincing story, and we are told it in the Bible not once but twice. Judges chapter 4 tells it in prose, as I have just summarized it. Chapter 5 tells it in verse, which Deborah composed and sang (with Barak providing a base or baritone descant). She thus emphasized a point which all sensible heroes or heroines learn: those who compose for posterity their own account of their deeds, and so get their version in first, are more likely to be remembered with all honor—a lesson made use of by many heroes, as we shall see, from Julius Caesar to Winston Churchill. Deborah sang the earliest version of her victory song on the evening after the battle. Just as she followed Moses's example as a judge, so her song echoed the chant of exultation he composed after Pharaoh and his host were drowned in the Red Sea (Exodus, chapter 15). On this occasion we are told that Miriam, another prophetess, and Aaron's sister, "took up her tambourine, and all the women followed her, dancing to the sound of tambourines," and Miriam sang to them this chorus:

> *Sing to the Lord, he has risen in triumph,*
> *The horse and the rider hurled in the sea.*

But Deborah, after the battle of Mount Tabor, was more than a tambourine girl. She was a successful ruler in war. As she put it:

> *Champions were there none,*
> *None left in Israel*

Until I, Deborah, arose,
Arose, a Mother in Israel.

Deborah's song is a more sophisticated piece of poetry than Moses's victory hymn. Much had happened since the days of the exodus, and the Israelites had honed their poetic gifts perhaps by illicit contact with the Canaanites and other more advanced peoples. There is some splendid detail in Deborah's song, and a pulsating rhythm in the battle scene, recalling the thunder of the chariots—and bitter taunts at the missing Israelite tribes who were not in the battle in their nation's hour of need. There is also pathos. Deborah tells the tale of Jael and Sisera in more detail than in the prose version and the description of Sisera's awe-inspiring death is dramatic, with the throb of sickening repetition. But Deborah adds a touching coda, describing the anxiety of Sisera's mother, worried by his late return, "peering through the lattices," watching the high road through the windows of his palace, and repeatedly asking her attendant princesses:

Why are his chariots so long in coming?
Why is the thunder of his wheels so delayed?

Deborah, an imaginative and clever woman, puts herself in the place of the tragic mother, and ends her victory paean on a humane note of sympathy for the stricken woman. The true hero always ends a battle by thinking of the slain, including the defeated enemy.

We have, then, this picture of Sisera's mother straining her eyes through the lattice. That is so characteristic of the Bible, a great book of true history, written as vivid literature, full of character sketches of the mighty and the small, and warm touches of humanity. There is nothing quite like it in the other writings of antiquity, until we come to the Greek drama of the fifth century BC. Homer has not the same intimate power.

Deborah's heroic epic is essentially about women—herself, Jael and Sisera's mother—and it left a huge impression on Israelite womenfolk, many of whom could read the Scriptures, some even helping to write them (the Book of Ruth is certainly by a woman). Late in Israelite history, about 300 BC, there appeared a sparklingly written account of an event that had occurred a century before, about the decapitation of an enemy general, Holofernes, by a beautiful and clever Israelite heroine, Judith. Scholars often argue that the Book of Judith, which is in the Apocrypha rather than the canonical Old Testament, is a romantic novel rather than a historical account. There was indeed a general called Holofernes, in the time of Ataxerxes Ochus, who reigned from 359 to 338 BC. But the Book of Judith ascribed to him the wrong nationality, and there are other errors and inconsistencies. But it seems to me more likely that Judith was an actual person, whose deeds struck the Israelites forcibly, since they recalled Deborah and Jael. Judith conflates these two separate war heroines into a single magnificent human being, who is, in addition, rich, wise and ravishingly beautiful. There is exaggeration here, plainly. But perhaps not as much as desiccated biblical text scholars think. The history of the Jews is peopled with remarkable creatures, especially women.

As it stands, the Book of Judith is one of the most beautiful and satisfying of the Scriptures, written in superb Hebrew. It has a feeling almost of voluptuous luxury about it. The ancient Near East has moved on decisively since the time of the Book of Judges. Everything is on a much bigger scale. In place of Sisera's nine hundred chariots of iron, the enemy king, Nebuchadnezzar, tells his general Holofernes, "take under your command professional trained troops of 120,000 infantry and 12,000 cavalry and march out against all the people of the West who have dared to disobey my commands." There is a tremendous amount of oriental boasting. The king says he will "vent my wrath." The whole land "will be suffocated by my army." The "dead will fill the valleys and every stream and river will be choked with corpses." Holofernes too is a boaster. Having put down all the disobedient peoples except the Israelites, he is warned by Achior of the Ammonites that Israel is a dangerous opponent, protected by a powerful god, and to beware of attacking them. Holofernes laughs him to scorn: "Who are you, Achior—you and your Ammonite mercenaries—to play the prophet and tell us what to do and what not to do? Our King is himself a God who will exert his power and wipe the Israelites off the face of the earth." He boasts: "Their mountains will be destroyed with blood and their plains filled with their dead." He then marched his army to the Israelite fortress town of Bethunia, cut off its food and water and prepared to starve it out. "Throughout the town there was deep dejection," the account continues, and Ozias the magistrate and other elders were denounced angrily by the people for not submitting to Holofernes. They agreed to surrender if God did not rescue them in five days.

At this point Judith sent her maid (evidently a formidable woman in her own right) to summon the elders to her to explain their cowardly conduct. They came. Why was she so powerful, like Deborah, and able to get the leading menfolk to obey her summons? She was rich. Her husband, Menasses, had died of sunstroke "while in the fields gathering the harvest of barley." He left her all his property, "gold and silver, male and female slaves, livestock and land." She was pious too. For three years she had lived as a widow, refusing all offers of remarriage, dressing herself in sackcloth, fasting every day except Sabbath eve, and living in a humble shelter erected on the roof of her palace. She was "both beautiful and attractive," and she had a reputation not only for piety but for wisdom too.

She rebuked the elders for their lack of trust in God, and said she had a plan to deliver the town from Holofernes. But it was secret, and she would not tell them in advance. All they had to do was to open the gates for her and her maid, and let her go to the enemy camp. She then recited a tremendous prayer, imploring God, "the God of the humble, the help of the poor, the supporter of the weak, the protector of the desperate, the deliverer of the hopeless," to give to her, "widow as I am, the strength to carry through my plan and shatter their guile with a woman's hand." She then removed her sackcloth and widow's weeds, washed, anointed herself with costly scents, did her hair, put on a headband and dressed in her gayest clothes, put on sandals and anklets, bracelets, rings and earrings, "and all her ornaments," and made herself "attractive enough to catch the eye of any man who might see her." She told her maid to put up a skin of wine and a flask of oil, and a bag filled with roasted grain, cakes of dried figs and the finest bread.

Then the gates were opened for her, the elders amazed by her startling appearance, and lady and maid, with their goodies, passed out of sight. When she reached the enemy lines, she was promptly arrested and questioned. But she had prayed to God to give her a deceitful tongue (being accustomed, as wife and widow, never to lie), and she told them a persuasive tale, that she was running away from her own people and going to Holofernes with vital military information. They were amazed by her beauty and intrigued by her tale, and she was taken immediately to the general's tent.

Holofernes was "resting on his bed under a mosquito-net of purple interwoven with gold, emeralds and precious stones." He received her graciously, captivated by her appearance, and she told him a long, circumstantial tale about how she would lead him and his army to Jerusalem "and I will set up your throne in the heart of the City." He was delighted: "In the whole wide world there is not a woman to compare with you for beauty of speech or shrewdness of speech." He took her to his inner tent, "where his silver was laid out," and ordered dinner. She remained in the camp three days, eating her own kosher food, and on the fourth day Holofernes, deciding the time had come to possess her, ordered a banquet for himself and her alone, without any of his officers and attended only by his personal servants. She, having put all the enemy's suspicions at rest, and won permission to enter and leave the camp freely, put on her finery and came to the general's tent. He was "beside himself with desire," and "shook with passion." She ate and drank with him, but from her own watered-down provisions. He drank a vast quantity of wine, "more indeed than he had ever drunk on any single day since he was born." When it grew late, the servants discreetly withdrew, closing the tent from outside: "Judith was left alone in the tent, with Holofernes sprawled on his bed, dead drunk." She prayed for strength, "went to the bed-rail beside Holofernes' head, took down his sword and grasped his hair." She "struck at his neck twice with all her might, and cut off his head." She "rolled the body off the bed and took the mosquito-net from its posts." Then she gave the head to her maid, who put it in the food bag. They were let out of the camp, as usual, and made their way back to Bethunia.

The consequence of this daring assassination was a psychological victory. On Judith's orders, the warriors of the town hung Holofernes' head from the battlements, and prepared for battle. The captains of the enemy host went to their general's tent to rouse him for battle, found him dead, his head gone, and the Israelite woman missing. A great shout of horror and dismay went up, panic spread through the camp, and the men began to desert in fear. Then the Israelites issued forth and began to attack, and the entire army was soon in terrified flight. "The looting of the camp went on for thirty days." They gave Judith Holofernes' tent, "with all his silver, couches, bowls and furniture," and she loaded it up on her wagons drawn by mules. Then, having safely stowed the booty, she led a dance of triumph, singing a song which begins

> *Strike up a song to my God with tambourines,*
> *Sing to the Lord with cymbals.*

The poem, which she wrote, tells the tale with a flourish, emphasizing the role of women. She took all her spoils to Jerusalem, purchasing with it sacrifices at the temple, and presenting the net from Holofernes' bed as a votive offering. Having purified herself, she "returned to Bethunia and lived on her estate." She refused all her many suitors, gave her maid her liberty, lived to be 105, full of fame and honor, and was buried alongside her husband. The story ends: "No one dared to threaten the Israelites again in Judith's lifetime, or for a long time after her death."

The role that women play in the Book of Judges and indeed in Jewish Scriptures as a whole sometimes raises ethical problems in an acute form. Can Jael's actions be seriously presented as heroic? Can even Judith's? Jael's brutal murder of an exhausted fugitive while he was asleep strikes us as odious. Expert biblical commentators mitigate it by explaining how seriously he had broken the hospitality rules, but might not they be nullified by Sisera's desperate plight? He foolishly believed he was among friends, and she reinforced his self-deception by her guile. She was in no danger from this fugitive so could not plead fear in her defense. It is hard to see her as a heroine at all. Yet Deborah predicted she would have the honor of accomplishing Sisera's end, and Jael is accorded heroic status in Deborah's chant of victory. Deborah herself is rightly accorded heroic status, but her implicit approval of Jael's horrifying act raises questions about her ethics. But then, the time is the second millennium BC, a Hobbesian world where sheer survival is the object of statesmanship, and to be exterminated, or to exterminate in turn, is the routine outcome of battle. Barak, presumably with Deborah's approval, or perhaps on her express orders, wiped out Sisera's soldiers: "the whole army was put to the sword and died; not one man was left alive." Sisera, naturally, was to be killed too. But Jael saved Barak the trouble, and to her was the honor of completing the extermination, and the praise for doing so. As Deborah's song says,

> *Blessed of all women be Jael,*
> *Blessed of all women in the tents.*

Jael was certainly a heroine to Israelites who read the Scriptures, and it is not to be wondered that the author of Judith made her a compound of Deborah's authority and wisdom and Jael's daring. Judith, however, was altogether a more admirable and heroic figure than Jael. True, she was deceitful like Jael, and had to tell a great deal of ingenious and cunning lies over four nerve-wracking days. God gave her the skill to spin her tales and the fortitude to play her part. She knew that one false step would mean a horrible death (after rape) and therefore courage was an essential part of her feat—courage of a high order too, multifaceted, for it had to include both the effrontery of playing the irresistible harlot (and a praying one too) and the decisive physical feat of decapitation. She did this in only two strokes too: this was the customary number for the professional executioner of the Tower of London in Tudor times, and they used an axe, an easier instrument for the purpose (only the specialist at Calais used, like Judith, a sword).

Judith's exploit appealed powerfully to painters, especially in the sixteenth and seventeenth centuries, looking for a dramatic incident with which to display their skill in the biblical-historical genre. She was beautiful too, and a virtuous woman of noble blood. So she developed a notable iconographic persona in Christian art. In 1598–1599, the great Caravaggio, whose talents were exactly suited to this theme, did a tremendous *Judith Beheading Holofernes,* now in the Palazzo Barberini, in Rome. His ultrarealism and theatrical chiaroscuro were deployed to depict the precise moment Judith severed the head from the body and the blood began to spout in jets all over the canvas scene. The Judith theme was taken up by his brilliant pupil, Orazio Gentileschi, and Orazio's tempestuous daughter, Artemisia, the first woman painter of the age. In 1611 they jointly painted *Judith and Her Maidservant with the Head of Holofernes*. This is a less sensational effort, after the horrid deed is done, with the maid stuffing the severed head into her bag. Artemisia's version, now in the Detroit Museum of Fine Art, is full of drama, however, stressing the danger and fear of Judith that they may be discovered—she listens anxiously for a sound of the camp stirring. But the blood is missing. However, in 1613 Artemisia was brutally raped by Agostino Tassi, the painter her father had entrusted to teach her. He was sentenced to eight months in prison, a punishment she felt was outrageously inadequate for his repellent betrayal of trust, and shortly afterward she returned to the subject of Judith. This time she painted the actual beheading. This overpowering work, now in the Uffizi, Florence (there is another version in the Capodimonte, Naples) is much closer to Caravaggio's effort. But in truth, it is superior, both in the handling of the blood and in the position and musculature of Judith in her fatal stroke. She is sawing the head off, using all her strength, most convincingly. Where Caravaggio's Judith is awkward, Artemisia's carries complete conviction. She is confidently enjoying her work. It is as though Artemisia had rehearsed the act repeatedly and discovered exactly how it was done—and imagined she was herself decapitating mercilessly the rapist Tassi.

What the stories of Deborah, Jael and Judith convey, and the iconography confirms, is the element of physical ruthlessness involved in many acts of heroism, especially when they are carried out by women. For these acts of violence are not, and never can be, routine. Their courage has to be, as Lady Macbeth puts it, "screwed up to the sticking pitch," and once thus heightened, explodes in action of reckless and heedless intensity. Heroism is usually loveless, and when performed by women its element of hate and inhumanity appears particularly savage. The hero, and still more the heroine, must be capable of atrocity.

The prominence of women in the Bible continues to surprise us. It is marked throughout the Book of Judges. Even when women are not performing heroically, they are often the mainsprings of the action. This is particularly true of Samson, the most famous of the Judges, whose story fills chapters 13 to 16 of the narrative. It is his mother, not his father, Manoah, who is determinant in the events surrounding his birth (rather like Mary in the case of Jesus Christ), who dedicates his life to the Nazirite sect and who names him Samson. Some historians have tried to place Samson as the equivalent of the Greek Herakles (the Roman Hercules), a wonder-performing strongman, and entirely mythical. But Samson is in every respect a Jewish hero and wholly rooted in history. Unlike any of the Greek myth figures, he is a real man, who is comic as well as tragic. This Janus-faced dualism of laughter and tears is characteristically Jewish, appearing very early in the scriptural record. Samson is its epitome.

As a Nazirite, Samson must never cut his hair. His life is to be devoted to serving the Lord, and has many other duties and inhibitions. But Samson is a flawed hero. He keeps his vow not to cut his hair, knowing this is the secret of his strength, which he values and exploits ruthlessly. But he disobeys all other laws and customs, and even his service to the Lord is whimsical, spasmodic and often absurd. He has an unbridled lust for women, and women of an exotic kind, temptresses and wayward. He laughs, scorning the Israelite doctrine of apartheid that forbids its men to fancy shiksas, or pagan girls. The first we hear of him as an adult is his abrupt order to his parents to get him a Philistine woman from Timnath as his wife. When they remonstrate he tells his father, "Get her for me, for she pleases me."

This sets the form of Samson's entire career: the urgent surrender, the strong passions. In search of the Timnath girl he fancies his way is barred by a lion which he tears to pieces with his bare hands. He later finds a swarm of bees making honey in the lion's body, and this gives him an idea for a riddle that he puts to the Timnite youths who attend his wedding feast, the prize to be outfits of clothing. His betrothed, under threats that she will be burned alive in her parents' house, coaxes the solution to the riddle out of him, tells it to her kinsfolk, and so he loses his bet. Furious, he slays thirty men in Ashkelon, and takes their clothes to pay the forfeit. Already with a reputation as a troublesome fellow, with a taste for tricks and black humor, Samson is denied his wife, who is given to another. He is offered her sister instead ("better than she") but rejects such solace: instead he declares a personal war on the entire Philistine people. He captures wild animals, ties torches to their tails and sets them loose in the standing corn, destroying the Philistine vineyards and olive groves too. In retaliation they burn alive his wife and her family. So he "smote them hip and thigh with great slaughter," and then went down to live in a cave in the Rock of Etam.

At this point, Samson's personal saga of lust and revenge gets swept up into the general Israelite resistance to their Philistine masters. Fearful of punishment, the Israelites hand him over to the Philistines, but he breaks the ropes that bind him and, using the jawbone of an ass, slaughters a thousand Philistine men. At this point he, like Moses and Deborah before him, breaks into song and sings the "Jawbone Victory Hymn." He is then thirsty, and tells the Lord in his characteristic intemperate manner to give him water: "Must I now die of thirst and fall into the hands of the uncircumcised?" So the Lord splits open a rock and creates for him the Spring of Lehi.

Samson is now a hero, or judge, lasting for twenty years in the anti-Philistine struggle. But there is nothing systematic or dependable about his leadership. He is eccentric in his behavior, idiosyncratic in his worship of Yahweh, fond of dangerous jokes and, above all, a womanizer. He seems to learn nothing from his experiences and falls repeatedly into traps set by sirens. He goes to Gaza, pounces on a whore, sleeps with her, arouses the anger of the locals, who surround the woman's house, but then gets up at midnight, seizes the doors of the city entrance, lifts them off their hinges, and carries them on his back "to the top of the hill east of Hebron." He then falls in love with the Philistine temptress Delilah, who is encouraged by their elders to lie with him and coax from him the secret of his incredible strength. She makes several attempts but is baffled by his weird sense of humor. Finally, he admits to her that it is his hair which empowers him. She then "lulled him to sleep on her knees," calls to a barber who "shaved off the seven locks of his hair," and then wakes him with her triumphant shout: "The Philistines are upon you, Samson!" And they are indeed: "They seized him, gouged out his eyes, and brought him down to Gaza. There they bound him with bronze fetters and set him to grinding corn in the prison."

At this point, Samson is transformed from a seriocomic figure into a wholly tragic one, and becomes a true hero. His hair grows again, unnoticed by his enemies, his strength returns—and his worship of God—and when the Philistines take him to their great feast in the Temple of Dagon to taunt him, he gets a little boy to guide him to the central pillars. Calling on God to give him the power, he pushes aside the pillars from their bases and brings the entire temple down, killing "the lords and all the people who were in it." The dead include, presumably, Delilah, and of course the little boy, as well as Samson himself. This ruthlessness in heroism makes Samson the first suicide-martyr-mass killer, adumbrating the suicide bombers of today's Middle East. It makes even more explicit and horrifying the links between heroism and a brutal unconcern for human life, whether guilty or innocent. Samson kills *all* the Philistines, including the innocent child who had befriended him, and many of those in the crowd who had nothing to do with his blinding.

Nonetheless, Samson was honored then—his brothers carried his body to his father's grave and buried him there—and thereafter by the Israelites, being a hero in the teeming biblical pantheon. One reason is that the episode, though clearly circulated originally in an oral version—it shows all the signs of such delivery—was later written down in tremendously powerful Hebrew prose (and poetry), so that the key phrases resonate and the characters, Delilah particularly, emerge with stunning impact. The fact that Delilah's ruthlessness matches his own, giving Samson a dramatically worthy adversary, enhances his tragedy. As described, Samson too is a human figure, not a mythical one, a real person with a character compounded of strength and tantalizing weakness, a mixture of cunning and folly, a loner despite his lust, a riddle teller and prankster, not really serious until he becomes wholly tragic.

The Jews loved Samson, and still do. They discounted his failings, which anyway could be used as virtuous warnings against temptation. Josephus, in his *Antiquities of the Jews,* gives a glowing portrait, showing him a man of exceptional goodness, except in his relations with women. He adds: "That he let himself be ensnared by a woman is the fault of his human nature which is liable to sin thus—but credit is due to him for his surpassing excellence in all the rest of his life." Josephus especially stresses his courage, the true mark of a hero-saint. This theme is taken up by the early Christian writers. Athanasius put him on a moral level with Samuel and David. St. Augustus likened him to Moses and Daniel. Some early writers compared him to Christ, in sacrificing his life for the cause and the truth. Clement used him as a powerful argument for chastity. All ranked him as a saint.

His literary and artistic appeal was even stronger, and more durable, than his religious message. Peter Abelard was the first to write a poem about him as a figure of tragedy. He wins the praise of Chaucer's monk, and of Boccaccio, who simply blames Delilah for his weakness. The sheer potency of her beauty and wickedness and her devilish tongue, wrapping itself around the hero's psychological muscles and rendering him powerless, appealed strongly, to artists especially. The iconography of Samson is enormous, at all periods of Christian art. I will note only three images. The first is by Rubens, showing Samson asleep over Delilah's knees, a painting (now in the National Gallery, London) only recently authenticated. This has a blond Delilah presented not so much as a temptress as a Philistine heroine, doing her duty to her people: the darker figure of the sleeping Samson is the intruder—a hint of anti-Semitism, perhaps? Rembrandt, by contrast, was devoted to the Samson image, doing him time and again in different postures. Two pictures, one with Delilah and the other threatening his father-in-law, both in the National Gallery of Berlin, are notable. But the most extraordinary canvas, in the Frankfurt Gallery, shows the actual instant of the blinding and Samson's almost superhuman effort to break free from his captors before the last of his strength ebbs away. This is the masterpiece of Rembrandt's dramatic style, or period, and throbs with violence. In my view, however, the most memorable pictorial commentary on the betrayal is the huge canvas painted by the Jewish master Solomon Salomon (1860–1927) in 1886, which hangs in the upper entrance hall of the Walker Art Gallery in Liverpool and dominates it. Here Samson is captive but not yet blinded, and the eye-line he flings at the exultant face of Delilah, feasting her eyes on his humiliation, is electric, a touch of genius.

Samson, however, reaches his apotheosis in the work of John Milton. *Samson Agonistes* is his finest poem, more troubling and painful even than the opening books of *Paradise Lost,* for Milton was himself blind when he wrote it, and his sense of the hero's deprivation of sight is complete and gives an overwhelming horror to the scene he paints in burning words. He introduces a new character, Harapha, the Philistine giant who visits the imprisoned Samson to taunt, but flees in fear. He also presents his hero as a Job figure, struggling with himself and his weakness, and with God. It is the finest thing Milton ever did, combining effects from Euripides, Sophocles and Aeschylus, and throwing a bridge which the readers can step over, between Hebrew, Greek, Christian, and English cultures. Milton's Samson has been described as presenting, from first line to last, a completely new and moving interpretation of the story. But of course by the time Milton introduces his hero, he is in his last phase, his repentance is complete, he has outgrown all his childish comedy and playfulness and he has become a martyr and a saint as well as a worthy hero.

The Bible will continue to inspire visual artists, musicians, and poets until the end of time, albeit—alas!—each succeeding generation now is less familiar with its contents, and its ringing cadences. And the Book of Judges will remain one of the most fruitful sources of the inspirational characters and incidents. For it manifestly comes from the heroic age. It deals with events which took place at the time of the *Iliad*. Troy was as real then as Gaza. Samson might have challenged Achilles to combat. And Deborah could have exchanged views with her fellow prophetess Cassandra.

Yet if the Book of Judges represents the classic time of heroes in Hebrew history, we must not forget the greatest hero of all, King David. Although Moses is, in one sense, the dominating biblical figure, the supreme lawgiver and recorder without whom the Old Testament would not have come into existence, David is the culminating hero of the narrative. All biblical events lead up to him, the creator of Hebrew Jerusalem and the founder of the United Kingdom of all twelve tribes. And after him, all is diminuendo and decline, to weakness and defeat, exile and troubled return, finally to martyrdom and diaspora. David is the culminating keystone of the entire biblical arch.

He was also an extraordinary amalgam of diverse gifts and qualities, a man of profound faith, with an intense personal relationship with God, but also weak and sinful, though always repentant in the end. His character is so complex that, although he is brilliantly described in great detail in three of the best-written books of the Old Testament (1 and 2 Samuel and 1 Chronicles), his personality eludes us. He is many things, and nothing; impossible to encapsulate. I live with David. In my London home the conservatory is dominated by a bronze replica of Verocchio's statue of David with his foot on the severed head of Goliath. And outside, in the garden, there is an early nineteenth-century replica, made in marble dust by a Florentine sculptor, of Donatello's famous bronze (now in the Bargello), perhaps the most remarkable statue ever carved and cast. But both show David in his most celebrated role as the teenage warrior killing the Philistine giant with his slingshot. And there are many other Davids. First the shepherd boy, summoned from his flocks to be chosen from his many brothers by Samuel, and anointed as a future king and leader of Israel. Then the young killer capable of slaughtering a lion and a bear as well as a Philistine giant. Then the young celebrity, welcomed at court and becoming the intimate of King Saul's handsome son Jonathan, but arousing the jealousy and eventually the bitter hatred of the paranoid king. Then the hunted fugitive, taking refuge in his Cave of Adullam, and attracting to him similar outlaws and detribalized fighters. Then the guerrilla leader and resistance captain, shadowing Saul's army as well as the Philistine host. Then the ruler of Hebron and parts of Judah, a king-in-waiting who eventually succeeds to the throne of Israel after Saul's defeat and suicide, and the reign of his undistinguished successor, Ish-bosheth.

King David brings together, in a powerful kingdom, all twelve of the tribes of Israel. He secures his own personal city, Jerusalem, and makes it the capital. He builds his palace there, and stockpiles material for a great temple, engaging Phoenician masters to design it, though he does not live to see its birth. What he does do is create a resilient state and administration, with a professional civil service keeping accurate official records, as in other advanced kingdoms. He associates Israel with the modern trading economy of the Phoenician coast, raises living standards, increasing exports of food and wool, leather and other raw materials, and imports of luxuries. He takes his people into the Iron Age, with mines and forges and modern weapons. He improves the roads and the safety of travelers and merchants. But his reign is troubled by horrifying disputes within his own sprawling family, partly the result of his own weakness and sin, his many wives and adulteries, his favoritism and irresponsibility toward his growing (and grown) children, partly the inevitable problem of a spirited ruler driven by pride, ambition and lust spiced by incest. David's reign is the epitome of personal monarchy with all its glories and miseries.

But on top of all this teeming saga of striking events, successes and failures in ruling, there is the emotional David and the artistic David. He loves God boldly, sometimes blindly and overwhelmingly, but he also loves women—and even men—with comparable abandon, sometimes in the face of God's wrath. He is abandoned, sensual almost to madness, obsessive and reckless. Also noble, tender, sensitive, hypertensive and melancholy. These characteristics are linked to his love of beauty and his creativity. He is a harp player of extraordinary skill, a musician who can mesmerize his audience. He is a poet, a master of that superb form of Hebrew verse we know as the Psalms. He compiled the first section of the Book of Psalms (1–41) and the fourth (90–106) and most of these he wrote himself. The 73 psalms ascribed to him personally are outpourings of his thoughts and passions, longings, guilt feelings, repentances, prayers and religious philosophy, directly springing from his relations with God and with other people. Thirteen of these poems, drawn directly from his personal experiences, which can be linked to specific passages in the first and second books of Samuel, are masterpieces of the genre, emotion recalled in tranquil or anguished reflection, here given not in order of composition but in the chronology of inspiration by event: 59, 56, 34, 142, 52, 54, 57, 7, 18, 32, 51, 3, 63. There is nothing like this from the ancient world, certainly not from the pen of a busy, often distracted king. And David sang his psalms, and conducted choirs and orchestras. He danced them too, when appropriate. He did not leave victory and joyous celebrations to maidens with their tambourines, but leaped and pranced himself at the head of the processions of celebrants, to the amazement and delight of his subjects. Was there ever such a man? Such a liberator and kingdom founder, such a mover and transformer and builder, who explored and exhausted every aspect of the human condition, both earthly and spiritual? No wonder, even after the great dispersal, all Jews who had the semblance of a claim were proud to point to David as their ancestor. Even Christ, the son of God, had his Davidic lineage proclaimed, at least by his followers and biographers.

No wonder too that David eludes portrayal in words. The authors of the books in which he appears never quite put all the disparate parts of his personality and record together. The David of the Old Testament lacks integrity; the bits and pieces are all described but the whole man is not there. He eludes us. Artists have been similarly baffled. They can show the boy-killer of Goliath (Caravaggio, in a moment of self-laceration, puts his own tortured image in the severed head David is bearing off in triumph), and the ruthless seducer of Bathsheba, sending off her husband to death in battle so his wife can be enjoyed in safety. But the man as a whole they do not even attempt. Artistic re-creation has its limits. Writers without end have sought to give us the life of David, and continue to do so. Perhaps the astonishing skills of a Shakespeare or a Goethe or a Tolstoy might have been equal to it; but these men of genius were too wise to try. So David remains the superhero of the ancient world, but also the unknown one.

2

EARTHSHAKERS: ALEXANDER THE GREAT AND JULIUS CAESAR

The heroes of the Bible were stalwarts of an obscure sect operating on a narrow, almost tribal stage. Even King David was no more than a petty king who made little impact on secular history. Their fame rests essentially on the worldwide spread of Judeo-Christian beliefs and the literary magic of the book which recounts their deeds: a book which has been read by more people than any other.

With Alexander and Caesar we come to the two principal actors of antiquity who operated in the theater of the entire known world and became prototypes of the heroic character for the next thousand years. They carved out vast empires for themselves and hammered their names into the history of the earth. Each was brave, highly intelligent, almost horrifically self-assured, whose ambitions knew no bounds. Also selfish, cruel, without scruple and fundamentally unlovable. But they were admired, inevitably, more perhaps than any other two men of their kind. They were giantlike, almost superhuman in every respect. What are we, at the beginning of the twenty-first century, with all our moral sensibilities, and our painfully acquired knowledge of human evil, to make of these two alarming fellow men?

Alexander the Great (356–323 BC) was the son of a tyrant. Philip II, king of Macedon (reigned 359–336 BC), was murdered at the age of forty-six, in circumstances which are still dark, and possibly with his son's connivance. He was a man of formidable achievements. Greek power and glory had been built on the culture of the commercial city-state and a combination of idealism and realism. Philip replaced the city-state with the nation-state formed from his own people of Macedonia, and the polarity of ideals and realism with a brutal realpolitik. He was highly creative. He built up a formidable professional administration with the capacity to raise money in vast quantities and to stretch its communications almost indefinitely. He also created a permanent professional army, based on the infantry phalanx, which was without equal in its own age, but also containing a well-trained cavalry arm, a vast siege train operating equipment never before seen, and a highly efficient commissariat. This army could go anywhere and do anything, and without it Alexander could have achieved little. It was the most magnificent inheritance an aspiring world conqueror could possibly hope for, and he also inherited Philip's ambition, having subdued all Greece, to seek revenge for the repeated invasions by Persia by launching a campaign of conquest in Asia.

If Alexander's father was a tyrant of genius, his mother, Olympia of Epirus, was passionate, ambitious, unscrupulous and violent, as well as a mystic. Whether she murdered her husband is doubtful. But after her son died, she pursued her own career as a ruler in Epirus and Macedonia, killing those who stood in her way. She was eventually cornered and executed by the families of those she had slaughtered, since the Macedonian authorities were not prepared to put to death themselves the mother of Alexander the Great.

Alexander, therefore, had a sinister parentage. But he was well educated. Philip chose as his son's tutor the philosopher-polymath Aristotle (384–322 BC), son of a doctor, who had worked under Plato and thus had links with Socrates himself. Alexander's years under Aristotle form one of the most intriguing relationships of all antiquity. It astonishes me that none of the great dramatists or novelists—Shakespeare, Goethe, Ibsen, Tolstoy—took it as a theme. Aristotle was bald, thin legged, had small eyes and a lisp—there is in the Vienna history of art museum a statue of him, perhaps done from life. He made up for his physical defects by fancy dressing and warned off criticisms by a talent for mockery. He was an awkward person to deal with, for, like many pedagogues, he could be sarcastic and severe. His sheer knowledge was encyclopedic, but he also created the basic vocabulary of philosophy, which we still use, and he had a genius for clarity, order, taxonomy and the logical progression from one subject to the next, in ascending order of difficulty. He was a born teacher, and it is hard to imagine an instructor more qualified to prepare a clever, ambitious young prince for world responsibilities.

It is a pity that Aristotle, who wrote so much about almost everything, did not leave us a few paragraphs about his most significant pupil—what he taught him, how well it was assimilated and how the prince struck him as a character. Such a portrait would be of enormous help in making up our minds about this world hero. For the truth is, we have virtually no firsthand information about him. This is not for want of any effort on Alexander's own part. He prepared his own historiography with care. He took with him on his wars one of Aristotle's cousins, Callisthenes, whose specific post was to write up the campaigns. Others close to him whom he encouraged to tell the story were one of his leading generals, Ptolemy, later ruler of Egypt; his boyhood friend Nearchus, later his admiral; another naval commander, Onesicritus; his architect Aristobolus; and Cleitarchus, also on the expedition. All of them did as they were expected, writing up the story from their leader's viewpoint. Cleitarchus wrote within thirteen years of Alexander's death, and Onesicritus was even speedier. But, without exception, none of these firsthand eyewitness accounts has survived. All were used by later authorities, some of whom were excellent historians, and whose texts have come down to us. But the earliest of them,

Diodorus of Sicily, lived three centuries later. He effectively rewrote Cleitarchus's account. Quintus Curtis, writing about 45 AD, four centuries later, had access to all the Greek eyewitness accounts, and part of his work has survived. Plutarch, working early in the second century AD, and Arrian, writing a generation later, also used the contemporary sources in works we can read. But of course it is not the same thing. Apart from a few fragments, no letters or documents from Alexander's day are known. There are coins, for what they are worth. Alexander had a court portraitist, Apelles, and a sculptor, Lysippos. He even had an official engraver to picture him on jewels. But none of their original work survives. Modern writers have to engage in a complex and difficult "search for Alexander," rather than settle down to a straightforward source-based biography.

It is likely that Alexander was handsome, blond, curly haired, slightly florid, above middle height, quick spoken and highly articulate. He rose early. His diet was spare. He learned to drink heavily at banquets or symposia, and unlike the Athenians, he did not water his wine but drank it neat, Macedonian style. But he was not a secret or solitary drinker. He was a superb rider and audacious hunter, especially of lion. He wrestled but otherwise did no athletics. But he was skilled with the sword and spear and expert at all forms of arms drill—his trade from boyhood. He loved bathing. He dressed to be seen: a lion helmet with a high crest, linen body armor and a big, flaunting cloak. He had various wives and was polite to women. His male friendships could be close. But he was not a passionate man. He slept alone.

He read Homer all his life and knew passages by heart. It was, to him, a Bible, a guide to heroic morality, a book of etiquette and a true adventure story. No one in his day (or for long after) had any doubt that what Homer described had actually happened. Alexander also believed he was descended from Herakles (or, as the Romans pronounced it, Hercules) the hero, sometimes treated as a god, whose "labors" made him the most popular of all Greek celebrities.

This mythical ancestry—very real to Alexander—had an important impact on his life, inspiring in him a spirit of emulation in courage and daring, and a longing to perform comparable marvels. Herakles' horrible death (he had himself burned in a pyre on Mount Oeta when he found the poison of Nessus taking hold of him) also gave him a certain fatalism, which made the risk of death unimportant to him. Aristotle certainly encouraged Alexander's love of Homer but tried to get him to distinguish the element of truth in history from mythical accretion, and to learn from nature. On this point his pupil was enthusiastic. When Aristotle returned to Athens after the death of King Philip, he founded a school and began to collect manuscripts, the prototype for all the great libraries of antiquity. He also assembled maps and objects to illustrate his lectures, especially on plants and animals. Alexander gave him the large sum of eight hundred talents to start the museum, and thereafter he ordered officials of his growing empire to send to Aristotle any portable objects of interest and to describe the scientific novelties they came across. Alexander's scientific and collecting instincts, inspired by Aristotle, were in turn to lead to Napoleon Bonaparte, in his 1799 expedition to Egypt, to take a one-hundred-strong team of scholars with him.

Alexander began to plan his war against Persia the moment he took over the throne, and he launched it immediately after his army and fleet were ready. One of the things Aristotle encouraged was the study of maps—he made a collection of the best available—and Alexander's ability to read a map and follow it expertly both inspired his enormous territorial ambitions and helped to realize them (here too Napoleon, a superb map reader, resembled him). Alexander believed, wrongly, that the great eastern ocean (the Pacific) was not far from the Indus Plain, and this was why he thought he could conquer the whole world. But in general he knew very well what he was doing in the eastern Mediterranean and western Asia, and he did it promptly. He crossed the Hellespont in 334 BC, and thereafter, for the next eleven years until his death in 323 BC, he was continually fighting and traveling. His career can be divided into seven phases. First, he won the battle of Granicus, near the Hellespont, thus enabling him to free the Greek cities of what is now the Turkish coast. Second, he destroyed the Persian fleet bases on the coasts of Phoenicia and Egypt, thus depriving his enemy of a naval arm and giving the Greeks complete command of the sea. Third, he defeated the main Persian army at the battle of Issus, near what is now Alexandretta. Fourth, he captured the fortress of Tyre, thought to be impregnable. Fifth, rejecting the offers of the Persian king, Darius, to divide his empire, he invaded Babylonia and destroyed the remaining Persian forces at the battle of Gaugamela on the Mesopotamian plain. Thereafter Darius was no more than a fugitive, and when he shortly died, Alexander succeeded him as king, treating any subsequent Persian hostilities as mere rebellion. He took the main imperial centers of Babylon, Susa, Persepolis and Ecbatana. From the Caspian to the Hindu Kush he experienced little opposition. But the conquest of Bactria and Sogdiana (Turkestan) was difficult and took three years. Sixth, he led an expedition into India to the Lower Indus and won the Hydaspes battle. He overran the Punjab. But at this point his army refused to go any farther. The return to Mesopotamia was difficult and horrific, but successfully accomplished. The seventh and last phase, occupying the last years of his life, was taken up by his plans to merge the Greek and Persian elites into one super race of warrior-conquerors, and in planning the conquest of Arabia. He was about to set out when he died of a fever after a series of terrific drinking banquets.

This is an astonishing story, quite unprecedented and never again even equaled in the age of horsepower. Apart from his actual battles and sieges, Alexander traveled over twenty thousand miles, much of it on foot in difficult mountain and desert terrain. How did he do it? The most important factor, as always with successful statesmen and men of action, was sheer willpower. He had preternatural self-confidence and persistence, the feeling that it was right to do what he planned and that he could certainly do it. There is no substitute for will. Alexander almost certainly believed not so much in his divine mission as in his divine power. His descent from Herakles was one pointer. His mother had given him others. In Egypt he undertook a dangerous and difficult mission to see the oracle in the Siwa Oasis in the Western Desert, to ask if he had, or would have, divine status. This seems odd to us but was not strange to a Greek of the fourth century BC, with a credulous worldview in which the difference between a god and a hero was narrow. The answer was reassuring, and eventually Alexander began to assume divine honors and to believe that he was the son of a god—possibly Zeus himself—rather than Philip. His self-confidence was continually reinforcing itself, both by success and by intimations of immortality.

But there were ten important practical reasons why Alexander succeeded. First, though he inherited a superb military machine, he continually improved it (by intensive training) and, above all, by *leading* it. He always marched with the men. He invariably led from the front. In his set battles, he was in the first rank of the cavalry charges that usually proved decisive. In sieges, he was under fire all the time, helping to operate the giant machines. Second, his battle leadership had a record written on his body. He was wounded nine times, nearly always in sieges—four wounds were superficial but they left interesting scars; four were serious, one nearly fatal, and after it he was no longer nimble enough to fight on foot, sticking to horse or chariot to move about the battlefield. When the men grumbled or threatened mutiny, he told them: "I have been hit by a sword, lance, dart, arrow and a catapult missile." He stripped and showed the scars. (Did this gesture pass into Persian folklore? When I interviewed the last shah of Iran, he took off his shirt and showed me the bullet marks left by an attempted assassination.)

Third, though Alexander fought on foot when not leading his cavalry, and led from the front, he seems to have been able to direct the battle and, if need be, change his tactics in midencounter. This was because Philip had created a good type of staff officer, which his son improved. Fourth, his battles were well planned thanks to his thoroughness in getting the best maps and his ability to read them. Fifth, he understood the technology of the day (thanks, again, probably to Aristotle), so his siege train was of the highest quality and was efficiently used. The ability to take a strongly defended city was crucial to his conquests. Sixth, he understood the importance of sea power and used it on a large scale and with skill. Whenever practical, his conquests were combined operations. Seventh, he grasped the vital role played by good, safe harbors. The pharaohs had ruled Egypt for three thousand years without building a proper oceangoing port on the Nile Delta. He set about constructing one as soon as he took over Egypt, and Alexandria still functions to this day. (He founded the great library there too, again inspired by Aristotle.) Eighth, he turned the whole expedition into a continual adventure by stressing comradeship. His cavalry were his "companions." His infantry "loved the King and he loved them." He usually knew them by name. Shakespeare picked up this point for his Henry V: "We few, we happy few," and Churchill for his Battle of Britain fighter pilots. Ninth, there were more material rewards. The booty secured during the conquests was colossal, and Alexander ensured that the men got a fair share of it and, equally important, that it was sent home safely to their families in Greece. (This again was a practice successfully followed by Napoleon.) Finally, from first to last, Alexander thought, decided and above all, moved swiftly. He appreciated the importance of speed in war, and the terrifying surprises speed made possible. His enemies were nearly always stunned and shocked by his arrival. He invented the blitzkrieg.

I have been describing Alexander essentially as he was in his twenties: the field commander and reckless conqueror. This is the heroic Alexander. As he entered his thirties, a different personality began to emerge. In some ways it was a statesman. In others a tyrant. Let us look at each in turn. As his empire expanded Alexander devoted increasing attention to the problem of how it could be ruled permanently. He came to the conclusion that it could not be done by Greeks alone, a fortiori by Macedonians alone. There were not enough of them, and too few could adapt anyway. So he introduced a policy of what we would call aculturalization. Not only did he appoint Persians, Egyptians and Indians to senior posts (this policy did not succeed; most Persian governors, for instance, had to be removed), he also set up training programs. He began with the military, enlisting twenty thousand Persian youths as cadets in his army, to be taught Macedonian drill, weapons training, tactics and discipline. But he wanted to go further and have joint administrative training so that senior officials, whether Greek or native, would use common terms and methods, counting, language and procedures. He showed himself willing to adopt native dress, jargon and customs when appropriate. In the far distance lay a vision of a multilingual, multicultural empire, based on power sharing and acquiring the permanency which rule by consent alone (in his view) could bring.

Well: that is indeed a modern notion, is it not? The Greeks were not ready for it, and the Macedonians hated it. They wanted to rule as they had conquered: by the sword. Moreover, they saw in Alexander's aculturalization policy not so much statesmanship and power sharing as a desire to seize a greater measure of power for himself, and power of a demeaning oriental kind. Greeks and Macedonians alike had a sense of freedom, and the dignity of man. Primitive and precarious it might be, but it was there. Persians and Egyptians did not possess it or even understand what it was. In Alexander's army the notion of freedom had been accentuated by the idea of participation in a common adventure and the status of "companion." Blending the cultures ran horribly counter to this military solidarity of brave men. And then, was not Alexander in danger of "going native" and being corrupted by power, of embracing the ways of oriental despots? Indeed he was. Given his successes, and the power victory thrust into his hands—power over countless millions—it would have been astonishing if he had not been corrupted.

The issue began to come to a head over Alexander's experiments with the Persian-oriental custom of prostrating in front of the monarch, what the Greeks called contemptuously *proskynesis*. And in the background to this was the rumor that the king intended to have himself deified. The result was a series of dreadful incidents which show the world conqueror in a brutal and uncivilized light. The first occurred in 328 or 327, during one of the symposia of dinner and drinking parties the king held with senior commanders and cronies. Cleitus, or "the Black," was a senior cavalry commander, in his late forties, who had saved Alexander's life at the battle of Granicus. He was of the old school and did not like what he called Alexander's "newfangled ideas." He had a sarcastic turn too. When both were drunk on the neat wine the king insisted on serving, they had an angry exchange, and Alexander suddenly seized a spear and killed Cleitus on the spot. This was the first time he had committed an outrageous act on a prominent Macedonian, and he took to his bed in a nervous collapse of remorse and self-disgust. Cleitus was evidently not popular among the young commanders and the murder was not at first held against him, or so it seems. Indeed his court historian, Callisthenes, Aristotle's nephew, took the trouble to persuade Alexander that he must put the episode behind him. But Alexander also fell afoul of the younger generation of commanders—his own—when he suddenly accused Philotas, a fine officer and commander of the Guard Corps, of conspiracy. He had criticized the king's adoption of oriental habits and plans for mingling the ruling classes, but there is no evidence of a plot. But Alexander had him tried "before the army" and got a conviction, the wretched man being sentenced to "execution by a volley of javelins." At least he got a trial of a sort. His father Parmenion, one of Philip's generals and originally second-in-command of the conquering army, was simply murdered. The death of Callisthenes was even more repellent. He had steered the king through the Cleitus crisis, but he objected strongly to prostration. When the king, after a drinking dinner, held a kind of dress rehearsal of how he expected his senior men to perform obeisance to him in the future, the historian, with all the rationalist arrogance of the Athenian intellectual elect, flatly refused. Shortly afterward, he was accused of complicity in the "Pages' Plot," a conspiracy by six teenage boys who had privileged access to the king in his quarters, to murder him in his sleep. This followed an incident in which one of them was flogged for stealing the king's quarry in a state hunt. All the pages were well connected and (it emerged) had family grievances against the king. They were savagely tortured and confessed, implicating Callisthenes. Alexander executed them all, and tortured the historian for good measure, though accounts vary about how he died. Some say he was starved and dragged around the camp in chains until he succumbed. Another version is that he was gradually cut to pieces: first his ears, nose and lips; then he was shut in cages with fierce dogs and a lion, until a merciful friend slipped him poison.

Instances of Alexander's cruel, irrational, violent and drunken behavior increased as he moved into his thirties, and it is plausible that he was becoming an alcoholic, and that alcohol was a factor in his death on June 10, 323 BC. On May 29 he had attended a banquet given by Medius, one of his officers, and we know the names of the seven or so men who attended it, some of whom were said to be in a conspiracy to poison him. Historians learned to be skeptical of poison reports before the nineteenth century, and this is no exception. Alexander was in an area (Babylonia) where, to this day, infections of all kinds, some still mysterious and many of which could then have been fatal, are encountered by visitors, especially those who are incautious about their eating and drinking habits. Eumenes, the king's secretary, kept a detailed diary, later published by a writer called Diodatus. The diaries from the king's last weeks were quoted by later sources. They show that Alexander's drinking bouts were becoming more frequent and serious and were followed by terrible hangovers, in which he took thirty-six hours to recover. There were five such bouts in the last months of his life. There is a story that after one of them he tried to kill himself by drowning in the River Tigris, but that he was prevented and rescued by his Iranian wife, Roxane, or "Little Star," daughter of the Sogdian Lord

Oxyontes. At all events, he died in his bed, of fever and weakness. It was not exactly a hero's death, and no doubt Alexander himself would have preferred to have died in battle. His approaching Arabian campaign would surely have seen him off anyway, from sunstroke or heat exhaustion.

Alexander became the prototype hero for the Hellenic age and many centuries after because of the extent of his achievement and his personal courage. He left his mark in many ways. He founded seventy cities, many named after him, in western Asia and Africa, with square, straight streets and gardens in the "modern" fashion. He thoroughly understood the art and technology of war in its most advanced form, but he had many other interests too: art, architecture, science, gastronomy, metallurgy, exploration, energy sources, and many other curiosities. He poked his nose into everything and missed nothing. He patronized actors and the theater, and artists of all kinds. Under his patronage the visual arts of the Greek city-states, disparate and competitive, were united in a common Hellenic style. His empire as a unity, under one man, was ephemeral, but the three Greek kingdoms into which it dissolved lasted hundreds of years. His conquests, in fact, made the subsequent Roman Empire possible, and it can be said that, as a result of Alexander's imperialism, the world took a giant step forward, to civilized unity and globalization. The destruction of life and property caused by his invasion was more than compensated for by the economic consequences of his wars. Alexander found in the Persian palaces truly prodigious quantities of gold and silver bullion, and by minting this into high-value coinages, and putting the results into circulation, financing the upsurge of trade produced by his new harbors and pirate-free seas, Alexander enormously increased the wealth and production of the entire area under his rule and beyond it. Here again he adumbrated the Roman age.

That is the positive side. Against this, Alexander's territorial greed, which was insatiable, and his love of war and the actual business of fighting—the passion of his life—set the worst possible example for all future generations, especially since his crimes against peace and humanity were gilded over by his martial glamour, heroism and quasi-divine status. Many evil and ambitious men took heart from his record. His favorite painter, Apelles, to whom he showed extraordinary generosity, giving him on one occasion his current *maîtresse en titre,* created what the Romans believed was a symbolic picture of the king in his chariot, pulling behind him a bound prisoner. The Romans said this captive was War, and that Alexander was the warrior who, by his decisive triumphs, had "captured"—i.e., ended—war itself. The picture has not survived of course, but the interpretation, indeed the symbolism, is specious. At his death he was planning another war against the Arabs, on the grounds that they had not sent him an embassy—a mere pretext, said his colleague Aristobolus: "In truth, he was never content with his conquests and he wanted to rule everybody." He was the first man to aim, realistically, at world rule, at any cost. He was a murderer, and in his battles a mass murderer, a lifelong criminal whose crime was the supreme one of war.

And was Julius Caesar, who lived a quarter millennium later (c. 100 BC–44 BC), any better, morally? As a hero he was on the same universal level as Alexander, the cynosure of the entire Greco-Roman world into which Caesar was born. But Caesar's fame, and example, as a permanent feature in the life of politics and statecraft, lasted far longer. Caesar effectively transformed the Roman Republic into an empire. Caesar refused the crown himself, but he made it possible for his designated successor, Octavian, his great-nephew whom he adopted as his son, to accept it under the title of Augustus Caesar. Thereafter the emperors were all Caesars, the name of the man having become synonymous with world authority. This nomenclature long outlasted the empire of the West, which endured for more than four hundred years, and the empire of Byzantium, which continued until the mid-fifteenth century, and as kaisers, tsars and the like, European emperors continued to call themselves after Caesar until the end of the First World War. The name survived in other ways. The Dark and Middle Ages, even the Renaissance, were overshadowed by the shattered glories of imperial Rome. In large parts of Europe any fortified hill or ruin whose origin was ancient and obscure became known locally as "Caesar's Camp" or "Caesar's Tump" or "Caesar's Wall." There are over twenty thousand such place-names west of the Rhine and south of the Alps. Caesar endures in another and important way. He was a man of colossal energy and farsighted cunning. He aimed to conquer posterity as well as the world he lived in, and he knew that to do this he must get in his own version of events in good time—and what better way to do it than to write it yourself? So, amid all his other activities and worries, in camp and in moments of hurried repose, he wrote, and wrote, and wrote. He finished seven books of the *Commentarii de Bello Gallico,* his account of his conquest of Gaul, one covering each year from 58 to 52 BC (his subordinate Hirtius completed the tale to 50 BC, after his general's death). He also contrived to produce three books called *Commentarii de Bello Civili,* the civil war he started and won. These books, written at speed and under pressure, have the supreme merits of simplicity and clarity. Cicero, greatest of all the masters of Latin prose, who heard Caesar speak in the Senate and read his work, praised warmly the correctness and precision of his Latin. Caesar's simplicity was deliberate. He wrote not to show off or astonish but to get his point across. He said you should "avoid an unfamiliar word as a ship avoids a reef." So his works were widely read and survived, being copied in endless manuscript chains across the Middle Ages, and then being among the first to be printed. They served as primary school texts for countless generations of children learning Latin. In the 1930s, I began my Latin prose studies with *Commentarii de Bello Gallico,* book one, chapter one, just as I began Latin verse with the first book of Virgil's *Aeneid.* The best-remembered of all Latin tags are Caesar's supposed last words, "*Et tu, Brute?*" addressed in agony, both physical and mental, to the young Brutus, son of Servilia, the mistress whom Caesar loved the best after Cleopatra. But according to Suetonius, in his *Life of the Divine Julius Caesar,* the great man, who had been well educated and spoke perfect Greek, used the language of culture to rebuke his favored junior: "*Kai Su, teknon?*"

Caesar was born about 100 BC from a noble family, which claimed descent from Venus, and so from Aeneas, the Trojan refugee who founded Rome. His father died when Caesar was a teenager, and the family was poor anyway, so Caesar, hugely ambitious, had to do it all for himself. He had a clever, encouraging mother. He suffered from epilepsy, then common. ("The usual complaint," as Plutarch says.) He became bald early and responded by cutting his remaining locks short and shaving off all his body hair. He had an obsession with cleanliness. His physique was good and he kept fit. He swam well, and his skill saved his life when he swam across the harbor in Alexandria. He was a superb and fearless rider: could gallop bareback with his hands folded behind him. His energy was colossal and he spent his life transmuting it into speed. He thought fast, decided promptly, above all was quick to move. On this point he excelled over even Alexander. He could notch up a hundred miles a day, whatever the conditions, and keep it up day after day. His ability to move fast the length and breadth of a vast theater of war was one key to his success, and the despair of his enemies. He was born to soldier. He loved the life of the camp. While celerity in action was his strength, he also possessed the patience to train his men with infinite care. The Roman legions were good before his time, but, after it, they formed the finest professional army of antiquity, better even than the Macedonian phalanxes. He said of his men: "These legions can tear down the sky." He trained his staff and senior commanders as meticulously as he trained the legionaries, and as a result could delegate authority with confidence. But he also led from the front, at the front when necessary, wearing a flaming red cloak so he could be seen by friend and enemy alike—just like Nelson with his stars and braid, a dangerous ploy. We know that he fought no less than fifty battles, almost exactly the same number as Napoleon, a dozen more than Wellington, and like these two fortunate commanders he was never seriously wounded. He looked after his men, saw that they were well fed, gave them brilliant little pep talks (like Montgomery), which he recorded in his books. Like Alexander, he saw that they got their booty to their families, and when they were too old to fight he settled them on farms. For Caesar, war was a career, and he turned it into a worthwhile career for his men.

Why did Caesar need to turn to war to make his way? By his day the republic was over four centuries old and had become not only the greatest power in the world but extremely complicated. The emancipation from the Etruscans dated from 509 BC. By 295 all Italy was Roman. From 264 to 206 BC Rome destroyed its chief international enemy, Carthage. By 241 Sicily was a Roman province. There followed Spain, North Africa, Macedon. By 129 BC all Greece and Anatolia were in the empire; by 58 BC, when Caesar came on the scene, eastern Turkey, Syria and Cyprus.

To run Rome, Italy and the vast expanding empire, the republic developed a wide range of elective officers operating under a highly specific law code. Rome was a class society: nobles, knights, plebs, merchants, "New Men" (*Novus homines*), etc. That was the warp, which determined who was eligible for what office, and whether he was likely to get it. The woof was family, and family connections. The family was all important, as it is in Italy today. Indeed the more I study Rome in the last days of the republic, the more it seems to resemble Italy in the early twenty-first century. In theory the law was inflexible and all deciding. In practice it yielded to family, and money; and money was made available by the corruption which family networks made possible. Money made it easier to run successfully for office, as in the United States today. But also useful was fame. And the surest way to acquire fame—and, incidentally, money too, often in huge quantities—was service in extending the empire.

Thanks to family connections, and judicious marriages (three in all), Caesar made his way slowly up the official hierarchy, but also made sure he saw military action, first in Cicelia and the province of Asia. To further his career he went to Rhodes to study rhetoric (in Greek of course), and when returning was captured by pirates. Caesar raised the money to pay his ransom, then a further sum with which he equipped a private navy, captured the pirates and crucified them all. In 74 BC he equipped a private army to fight the king of Pontus. His breakthrough came in 61 BC when he secured the governorship of Further Spain, using borrowed money and family influences. He was now deeply in debt, but the governorship meant fighting, and fighting meant loot. Caesar was so successful that he returned rich, paid all his debts and had cash to finance further moves. He was given charge of Cisalpine Gaul, and by a stroke of luck he soon got Transalpine Gaul too.

The conquest of Gaul, 58 to 50 BC, was the fulcrum of Caesar's life. It added an enormous and, potentially or actually, rich territory to the empire, and made Caesar, in cash terms, the richest man in Rome. It was done without any authority from the Senate, indeed against the rule that no further territory was to be conquered without specific permission. The cool effrontery with which Caesar describes his step-by-step conquest, never using the word "I" or "my"—always "Caesar" or "Caesar's"—is one of the most consummate impostures in literary history, let alone political history. He makes everything he did the mere inevitable response to events. The Caesar he presents was guided purely by his devotion to Rome and its interests. Personal ambition did not come into it at any stage. Everything was the ineluctable logic of history, fate and geopolitics. The simple, terse and factual prose adds greatly to the plausibility of his tale. The Caesar who emerges is not only superbly capable and efficient but also wise and moral, even innocent. The work says, "This is what Caesar had to do."

It may be that this is what he really believed, and that Caesar had powers of self-deception, or the ability to identify the republic's interests with his own, of a rare order. Caesar had a wonderfully optimistic temperament. One envies the delight he took in soldiering. His battle cry was "*Felicitas!*" He was in truth Wordsworth's "happy warrior, which every man in arms would wish to be." He led charges "at an enthusiastic gallop." He drilled his troops "not like a commander of veterans but like a gladiatorial fencing-master with new recruits." It was an art, a culture, he was imparting, with the delight of a virtuoso. On campaign in Gaul, he was always cheerful and self-confident. His approach to difficulty was: all problems are soluble. When fodder ran out he fed his horses on seaweed washed in fresh water. There were a lot of jokes and bawdy songs in his army. The soldiers called him Baldy or "the Bald Adulterer." They joked about his supposedly, when a youth, staining the bed of King Nicomedes of Bithynia (a rumor Caesar always hotly denied). It is impossible to deny that Caesar created and commanded not only a superb army but a contented one, a willing and smiling instrument.

Yet we have to say that the eight-year conquest of Gaul was, from an ethical viewpoint, one of the great crimes of history. And Caesar could not claim, like Alexander in subduing the Persian empire, that he was ending an oriental tyranny and replacing it with Greek civilization, including notions of freedom. Rather, he was stamping on tribal kingdoms where ideas of freedom were already living, and reducing them to the slavish mold of a colony. His expedition to Britain, flowing logically from his Gallic conquests, indicates he would have done the same there, had the timetable of his ambition permitted. The human cost was enormous. Caesar, in organizing his "triumphs" in Rome after his victories, always gave accurate figures of the numbers slain, based on body counts and likely, therefore, to be underestimates. A figure compiled at the time gives 1,192,000 slain, a large majority Gauls. This does not include prisoners, naturally. Their fate could be gruesome. Vercingetorix, who led the Gaulish rebellion against the conquest, was kept in chains for six years, then put to death. Caesar's triumphant celebrations, conducted on an unprecedented scale to curry favor with the Roman plebs and demoralize his rivals and critics, involved real battles between his prisoners and criminals under sentence of death—and the prisoners might also be savaged by wild beasts or trampled to death by elephants.

The complex maneuverings which followed Caesar's success in Gaul and culminated in his unlawful incursion into Italy at the head of his army—"crossing the Rubicon" (the river which formed the provincial boundary)—can be seen as high politics or mere gang warfare, struggles between rival mafias for jobs, offices, the chance to run systems of corruption, and the right to inflict judicial murder on your opponents. It led inevitably to large-scale civil war; the battle of gangs for territory was fought on an intercontinental scale. Caesar campaigned swiftly and efficiently in Italy and Greece, in Spain and Africa, by land and sea. Throughout the history of Europe, princes and future kings would be taught about Caesar's battles—all of which he won—in order to learn about war and ruling. But these struggles can be seen in the harsh light of gangster criminality, as well as the soft glow of historic grandeur. Pompey was a major hero figure like Caesar, the two men sharing a hatred for pirates, and Pompey rid the Mediterranean of this scourge as soon as he had the power and chance. But the empire, though vast, was not wide enough for them both. Caesar, one of history's greatest generals, and with an army he had personally trained and commanded for a decade, inevitably prevailed. Pompey fled to Egypt, arriving three days ahead of Caesar's ships. The Greek king, Ptolemy XIII, decided to win the great man's favor by cutting off Pompey's head, and presented it when Caesar arrived in the palace. Caesar turned away in sorrow and disgust and wept when the dead man's signet ring was handed to him. He had wanted to pardon his beaten opponent. So he gave him a decent burial and was generous to his remaining followers.

Caesar's sojourn in Egypt in 48 to 47 BC forms a lurid and dramatic episode, the stuff of plays, as it has often become. He put up at the palace, and there Ptolemy's elder sister Cleopatra, then twenty-one, arrived in a small boat and was delivered to him in a bag. Plutarch says "her appearance was majestic and pitiful." She won Caesar's heart. Suetonius says that of all "foreign women," she was the one he loved most, and longest. She was well educated and spoke many languages. A book on cosmetics was named after her—perhaps she wrote it. Her great charm was her thrilling voice: "It was a delight to hear its tone—her tongue was a lyre of many strings."

Although Cleopatra became Caesar's much-loved mistress, the ministers of her young brother stirred up the mob to attack the Romans (Alexandria now had a population of 500,000, and was notorious for its anti-Semitism and xenophobia) and Caesar was in effect besieged in the water palace and the island lighthouse of Pharos. Caesar set fire to the Egyptian warships in the harbor, and this was the occasion when he saved his life by swimming out to his own ships. The fire did great damage and, it is reported by Plutarch, destroyed the great library of manuscripts, the largest in the world (the total of volumes, or rolls, is variously given as 100,000, 400,000, and 700,000). This was founded, on Aristotle's model, by Ptolemy I, Alexander's immediate successor, and its loss was a blow to scholarship still felt to this day.

Caesar returned from Syria with reinforcements, destroyed all opposition in Egypt and installed Cleopatra as sole ruler (her brother was drowned in the Nile). He went quickly to Asia Minor, where he defeated Pharnaces of Bosphorus at Zela. This was the occasion when he reported laconically: "*Veni, vidi, vici.*" He returned to Africa, via Rome, won the battle of Thapsus, went to Spain again, and finally destroyed the remnants of the opposition at Munda. This was said to be his hardest battle, and on the whole Caesar had to do it all himself, in a hurry. Once the civil war began, he never had time to build a reliable command structure. Antony was loyal, willing and eager, but by temperament a Number Two. Tall and handsome, according to Plutarch, "he had a well-shaped beard, wide forehead and curved nose which made him look manly and rather like the statues and paintings of Hercules." Good-natured, lascivious, a boaster and hard drinker, he did good service under Caesar but needed orders. As Plutarch put it, "Cleopatra found Antony already tamed."

Back in Rome, the civil war over, Caesar was made dictator with the mandate to hold elections for offices. In 46 he became dictator for ten years, and in 44 for life. His last years in Rome saw him inaugurate a series of fundamental reforms, in citizenship, taxation, land settlement, municipal corporations, colonies and the Senate (which he enlarged to nine hundred). He also prepared to go on campaign in Asia against the unconquered Dacians and Parthians. He refused the title rex, but accepted many other unprecedented honors, which the stiffer element among the republicans could not bear. Caesar tried to reassure them by dismissing his Spanish bodyguards, but that gave his hard-line opponents the opportunity to carry out his murder. He was stabbed to death in a Mafia-style killing in the Senate on the ides of March 44, and died at the foot of Pompey's statue. This outrageous crime in a sacrosanct place, designed for debate and the enactment of law, was hailed by Cicero, supposedly defender of the republic's highest standards of constitutional behavior, with delight. The truth is, the republic was a sham, the rule of law a fantasy, and family, money and, in the last resort, force ruled Rome and its empire. Caesar's entire career essentially destroyed nothing, but exposed everything.

There is no doubt that Caesar was a man of exceptional ability over a huge range of activities. Among his qualities Pliny the Elder listed as foremost: great mental power, energy, steadfastness, a gift for understanding "everything under the sun," vitality and "fiery quickness of mind." Few men have such a combination of boldness, shrewdness and wisdom. But if he was the model of kingship for two millennia, he was also the inspiration for dictators and tyrants. Napoleon, at the height of his power, when all Europe danced attendance at his levees, said to Goethe: "You should write about the death of Caesar in a worthy manner, better than Voltaire. It could be the greatest task of your life. The world should be shown how happy Caesar would have made it, how much better everything would have been if only he had been given time to bring his sublime plans to fruition." That is one view. Another might ask: "How many more would Caesar, had he lived, have killed." Napoleon killed five times as many as Caesar's total, perhaps five million. Mao Tse-tung, another admirer of Caesar, killed seventy million. These things need to be weighed when we tell the stories of heroes.

3

FEMINIST FIRE AND SLAUGHTER: BOUDICA

Boudica is one of the most striking figures of antiquity and the first heroine of Britain's history, rich in fiery queens. She was tall; had red hair down to her waist; was commanding, brave and a notable orator. The Romans said her voice was harsh, but any woman seeking to establish authority over an assembly of men is open to this accusation, as Margaret Thatcher found in the House of Commons. Boudica was a contemporary of the emperors Claudius and Nero. She led a surprisingly successful British revolt against Roman rule in AD 60–61, that is, at the time when St. Paul was writing his epistles to the Corinthians and St. Mark composed his gospel.

Her name raises a problem. So do the names of other ancient Britons. Just before the Roman conquest of Britain, which began in AD 43, the long-reigning chief king of the Britons was Ceurobeliuus, who appears as Cymbeline in Shakespeare's play. His son Caradoc, the first man to play a notable role in the history of Britain, is also known as Caradog or Caratacus or Caractacus. Tacitus, one of our two sources for Boudica and her time, misspelled her name with two *c*'s instead of one. In the Middle Ages, copyists compounded the error when the *u* was replaced by an *a* and the second *c* by an *e*. This is how we ended up with Boadicea, the name by which she was known until recently. Even today there is confusion, since adults like me pronounce it Boudica with the accent on the middle syllable, whereas schoolchildren, by what authority I know not, are firmly told to put the accent on the first. But what's in a name? Tacitus is one of the best and most famous of Roman writers, and hard and bitter were my struggles, as a twelve-year-old, to master the elliptical Latin of his *Annales.* We know his middle name was Cornelius, but his first name may have been Publius or Gaius. We are unsure of both his date of birth and his death date. These are reminders of how many lacunae there are in our knowledge of such distant times.

We have three literary sources for Boudica, two by Tacitus. In *Agricola,* the life of his father-in-law, later governor of Britain, which was written within living memory of the revolt, Tacitus says that while the then governor, Suetonius Paulinus, was conquering Anglesey, stronghold of the Druids, the oppressed Britons ("the whole island") rose "under the leadership of Boudicca, a lady of royal descent—for Britons make no distinction of sex in their leaders." Tacitus's second account of Boudica, in his *Annales*, is fuller. It says that Prasutagus, king of the Iceni, a rich man, had made the Emperor Nero coheir with his two daughters, hoping thereby to preserve his kingdom and family fortune. But his will was ignored, his widow Boudicca (again misspelled) was flogged, and his two daughters raped. The dead king's property and estates were seized by Roman officers and his family treated like slaves. As a result, says Tacitus, the Iceni rose in revolt, backed by the Trinobantes, who had grievances of their own, and other tribes. They destroyed the colony at Colchester, which was unwalled, annihilated "the ninth Roman legion," which tried to relieve the town, and forced Governor Paulinus to evacuate London, which was also destroyed. He adds that "Verulamium [Saint Albans] suffered the same fate."

Contradicting his account in *Agricola,* which claims that the Britons stormed Roman forts, Tacitus says that "forts and garrisons were by-passed," the Britons going "for where loot was richest and protection weakest." He says 70,000 Romans were killed, explaining: "The Britons did not take or sell prisoners. . . . They could not wait to cut throats, hang, burn and crucify—as though avenging, in advance, the retribution that was on its way." In response, Paulinus collected 10,000 troops and lured the Britons into a pitched battle, on grounds of his choosing, at a place Tacitus does not identify. Modern topographical experts have examined various possibilities: Paulesbury near Towcester in Northamptonshire, Wroxeter, Mancetter in Warwickshire (not far from Bosworth where Richard III met his fate at the hands of Henry Tudor) and one or two other places. But all are unsatisfactory for one reason or another, and none has been corroborated by physical finds on the spot. Tacitus says the Britons congregated in such numbers, on foot and horseback, and "their confidence was such that they brought their wives with them to see the victory, installing them in carts stationed at the edges of the battlefield." He says that Boudica "drove round all the tribes in a chariot with her daughters in front of her, and addressed to them a fighting speech with marked feminist overtones, showing her bruised body and outraged daughters, and concluding: 'You will win this battle or perish. That is what I, a woman, plan to do. Let the men live in slavery if they will.'"

Tacitus says in the *Annales* that the outcome was an easy Roman victory over a British army of which more than half were women. He says 80,000 Britons fell at a cost of 400 Roman dead and "a slightly larger number of wounded." Boudica, he says, "poisoned herself." The rebellion ended in a famine among the Britons.

These accounts are supplemented by Cassius Dio's record of the invasion and occupation of Britain, and its aftermath, written in Greek about 150 years later, but based on works which have not survived, by writers who knew more about the events than Tacitus. Dio describes the revolt as "a terrible disaster." He puts the numbers of Romans slain at 80,000. Two cities were sacked and "the island lost." He comments: "Moreover all this ruin was brought upon the

Romans by a woman, a fact which in itself caused them the greatest shame." Dio gives a more plausible account of the socioeconomic causes of the revolt than Tacitus. He says one factor was the recalling of loans originally made to prominent Britons in the time of Claudius. He blames the chief minister and stoic philosopher Seneca, man of enormous wealth who, after Boudica's death, was forced to commit suicide, in AD 65, for political offenses. Whether we accept Tacitus's or Dio's explanation, it is clear that the Roman occupation was marked by brutal financial exploitation of the ruling elites, and oppression that hit the natives of all degrees. Dio says that Buduica (or Budhika), as he calls her, was chosen leader by the tribes and "directed the conduct of the entire war." He says she had "greater intelligence than is generally found in women," was "very tall, in appearance most terrifying, in the glance of her eye most fierce, and her voice was harsh. A great mass of the tawniest hair fell to her hips. Around her neck was a large golden necklace. She wore a tunic of many colours over which a thick mantle was fastened by a brooch . . . [She] grasped a spear to help her terrify all who saw her."

The speech Dio ascribed to Boudica in spurring on her followers is much longer than the Tacitus version but follows the same lines: freedom or death—better to perish in battle than live under Roman rule as slaves. It contains an added note: Boudica stressed that the Britons were a special people, separated from the rest of mankind by a sea, and enjoying, until the Romans came, a liberty unknown elsewhere. According to Dio, she practiced divination and magic, concealing a hare in her garments, which, at a key point in her speech, she let escape to see how it would run. Dio shows her displaying considerable knowledge of Mediterranean deities, and explaining "we have by now gained much learning from the Romans."

Much of what Dio writes seems to us fantastic. He describes obscene cruelties inflicted by the Britons on Roman women, for example—cutting off their breasts and sewing them onto their mouths. He says that Boudica's army totaled 230,000. But his description of the final battle, more detailed than the Tacitus account, is more convincing. He says it was very close, and many Britons escaped and were preparing further resistance. But "Boudica fell sick and died." The Britons "mourned her deeply and gave her a costly burial. But feeling that now at last they were really defeated, they scattered to their homes."

It can be argued that Boudica posthumously gained her point. Roman administration improved markedly in honesty and acceptability. Indeed, it was popular, and Britain became prosperous. A Roman-Briton ruling class was most reluctant to see the Romans leave. Indeed, in the only other ancient reference to Boudica, by the British author Gildas writing in the sixth century after the Roman withdrawal and the beginning of the Anglo-Saxon conquest, the Romans are presented as "goodies" and those Britons who resisted them as "baddies." He says: "A treacherous lioness butchered the governors who had been left to rule." This is clearly a reference to the warrior queen and, since Gildas had not read Tacitus and Dio, suggests that her folklore memory was still strong half a millennium after her death, though now reflecting the comforts enjoyed under a Roman rule seen as benevolent.

The archeological evidence is both firm and disappointing. First, because it shows beyond doubt that disasters occurred in Colchester, London and (to a lesser extent) in Saint Albans at a level which fits the literary evidence. It shows these three towns were burned. They were made chiefly of wood, and what archeologists call a "destruction layer," thirty to sixty centimeters thick and black or red in color, confirms the towns were put to the torch about AD 60. But specific physical evidence of the revolt and of Boudica herself does not exist, or has not so far been found, despite strenuous efforts that still continue. There are theories rather than proofs. In the Museum of London there are skulls from the Walbrook stream, said to be those of her victims, but they cannot be precisely dated, and skulls are found there from various periods. At Colchester there is a tombstone of a Roman cavalryman, Longinus Sdapeze, said to have been mutilated by the rebels. But recent research suggests rather that the damage was done by archeologists in the 1920s. A large enclosure discovered at Thetford has been called "the palace of Boudica." But similar structures exist elsewhere and their purpose is debatable. No coins, either of Boudica or her husband or anyone who can be connected to them, have been identified.

Yet Boudica is an immensely striking and even attractive figure—and national concept—and, in the absence of real evidence, imaginations have worked hard. In the early sixteenth century, Polydor Vergil, an Italian immigrant who published the first edition of Gildas and a history of Britain, presented "Voadicia" as a Northumbrian lady and had her burning down Doncaster. His contemporary Hector Boece made her a Scottish heroine from Falkirk in his *Chronicle of Scotland*. The Elizabethan historian William Camden, while relocating her correctly, thought he had found her coinage, but they turned out to have been issued by Bodvoc, a Gloucester magnate. In the years 1609–1614, the playwright John Fletcher wrote a play called *Bouduca,* which cunningly surrounded her with Druids, Caractacus and other bits of ancient British furniture, though to please James I he made her a witch and a horrible woman. This play was astonishingly successful and was revived over and over in the seventeenth, eighteenth, and nineteenth centuries. It is odd that Benjamin Britten, who hated women so much, did not turn it into an opera. Fifteen years after Webster's play was first produced, Edmund Bolton produced the theory that Stonehenge, whose purpose and date had baffled antiquarians, was in fact Boudica's tomb—had not Dio Cassius written of her "costly burial"? As late as the time of the French Revolution, Edward Barnard, in his *Comprehensive, Impartial and Complete History of England* (1790), stated: "Stonehenge was erected as a monument to commemorate the heroism of Boadicea." Barnard, however, admitted that an alternative burial place was Parliament Hill Fields in London, though this turned out to be a much earlier Bronze Age site when excavated a century later. Another site favored in the eighteenth century was Gop Hill in Flintshire, where locals said they had seen her ghost driving a chariot. There is another theory, held by the people who congregate for the summer solstice at Glastonbury, that she is buried deep below platform 8 at King's Cross Station in London.

Interest in Boudica increased in the late nineteenth century pari passu with the belief that Britain's unwritten constitution was of immemorial antiquity, and that she had played some part in its foundation. She ceased to be a baddie, and became an undisputed heroine forever. In the age of Gladstone, who encouraged the project, the sculptor Thomas Thornycroft worked on a massive presentation of the queen in her chariot, and her two daughters. Queen Victoria, who thought her treatment by the Romans "outrageous" (she too had been widowed early and had many daughters) was particularly keen that Boudica should be given a fine memorial, near the Houses of Parliament. Before his death, Prince Albert had urged that Boudica's "regality" should be stressed and that the chariot should be a "throne upon wheels." Thornycroft began work in 1856 during the Crimean War, but the statue was unfinished at his death in 1885. It was then cast in bronze, presented to the nation in 1896, and finally put in place near Westminster Bridge in 1902. It is a splendid piece of work, exactly what monumental sculpture should be. Children love it. So do feminists. Placed as it was, it inspired the suffragettes in the years 1910–1914, anxious as they were to engage in highly publicized activities near Parliament. It is a fitting memorial to a woman part fact, part fiction, who has inspired countless drawings, prints and paintings, and various plays and novels. As we have seen, not much is really known about her, and the public knows even less. But she rings a kind of muffled bell in all British, and many American, hearts, and is liable to come to vigorous life again whenever her tragedy strikes a topical note.

4

EXEMPLARY HEROES: HENRY V AND JOAN OF ARC

The pagan classical world had an empirical morality which celebrated the skillful and successful use of force. It feasted and immortalized those who were able to wield it. It averted its eyes from failure and regarded the weak and helpless with indifference. The Judeo-Christian tradition was quite different. Although the Israelites in the Davidic kingdom were briefly strong in their part of west Asia, David himself in his defining struggle with Goliath epitomized the heroically puny, and for most of their existence the Israelites were struggling against great empires, often unsuccessfully. The Psalms are usually the poetry of the weak, the abandoned, the helpless, the forsaken. The Israel of the Maccabees was a resistance movement against their Greek overlords, inheritors of Alexander's empire, and those killed in the struggle were treated by the Jews as hallowed saints. Thus the concept of the martyr was born. It was seized upon eagerly by the first Christians, beginning with St. Stephen, stoned to death, professing his faith in Christ in his last moments, embracing his fate calmly, unprotestingly, almost eagerly.

All the heroes of early Christianity were martyrs. St. Peter crucified, on his insistence upside down so as not to compete with

Christ, who was upright. St. Paul, beheaded. St. Lawrence, roasted to death by a gridiron. St. Sebastian, sentenced to be shot by archers but surviving, and then being battered to death with cudgels. And then the brave women: St. Barbara, first shut in a tower (her symbol), then tortured to death. St. Catherine, whom the pagans sought to break on the wheel (her symbol) but it was the wheel that broke; and finally beheaded. St. Agnes, a mere child of twelve when she was martyred by being stabbed in the throat. And so on: a long litany of glorious horrors, the blood of the martyrs watering the seeds of the church until, in God's good time, and in the reign of Constantine, the state itself became Christian and the faithful emerged from the catacombs to inherit the land.

Christian heroism, then, was innocent, suffering, almost pacifist, in its earliest phase. Next to the life of Christ and the Virgin, Christian art, as it burgeoned in the Middle Ages and climaxed in the Renaissance, concentrated on depicting the martyrs in their sacred death agonies. The favorite heroes were those who met the most gruesome deaths, or those which could be rendered by artists with realistic and edifying relish: the archetype of all pictorial martyrology being, perhaps, the vast canvas of St. Sebastian being shot by the archers, the masterpiece of the Pollaiuolo brothers, now in the National Gallery, London.

But in a man's world it was unlikely that the dominant culture, as Christianity became in Europe, should confine heroism to unresisting suffering and the passivity of women. Sluggishly to begin with, then with growing power, Christianity responded to the threat of Muslim conquest, especially in Spain. In the struggle to turn back Islam, the concept of Christian knighthood was born, and quickly attracted its own saints, real or imaginary, male, in armor, with sword, lance and shield, horsed and triumphant, St. George being the archetype. This new kind of hero, from the end of the eleventh century, assumed the mantle of the Crusades, with a white surplice over his armor, the red cross of Christ emblazoned on it.

The Christian hero was not merely a man of physical

characteristics—courage, skill at arms, and a commanding presence on the battlefield—but sought to display metaphysical qualities too. These included religious faith and devotion, which took the form of founding churches, abbeys and nunneries with the spoils of battle, and the ideas brought together in the concept of chivalry: courtly manners, hospitality, patronage of minstrels, poets and artists; above all, respect for women. The virtues thus embodied in the earlier spirit of martyrdom were transmitted and reborn in knightly conduct which was gentle as well as bold, delicate and thoughtful toward others as well as resolute in righteous conflict. He was, above all, honorable. Christian kings became ex officio heads of the knightly profession, and modeled themselves on earlier prototypes: King Arthur in England, Charlemagne in France, Otto the Great in Germany. The king was the fountain of honor, and to institutionalize the role, kings founded orders of chivalry, such as the Order of St. Denis in France and, above all, the Order of the Garter in England. This was established in the mid-fourteenth century by Edward III, himself a warrior of great experience and success. It had a round table where the knights feasted at Windsor Castle, consciously re-creating the Round Table of knights who gathered around the semi-mythical King Arthur, "the last of the Romans," who defended Christian Britain from Saxon pagans.

It is fitting that the exemplary medieval hero should come from the ranks of kings, and no one fits the part more completely than Henry of Monmouth, king of England as Henry V. He is a true hero in many ways, and has a strong claim to be rated the greatest of all English monarchs. I thought it would be instructive to compare and contrast him with Joan of Arc, La Pucelle, who occupies a comparable position in the French medieval pantheon. Joan was only a small child when Henry died, but they were pitted against each other historically in the great Hundred Years War between the French and English crowns—Henry establishing a unique position of power in that long conflict, and Joan, the unique woman warrior of medieval chivalry, playing a decisive role in overthrowing Henry's work, under his feeble heir.

A medieval king of England did an executive job. He was head of state and head of government, head of the armed forces, leader ex officio of the nobility and the landed interest, the active patron of the mercantile class and ultimate source of justice for the peasants, the protector of the church, as well as its chief source of appointments, and not least the presiding officer of the central court, where law was made as well as enforced. The king took all important decisions, or delegated them at his peril, and he was constantly on the move, to be at the scene of the action.

Medieval kingship wore a man out, or killed him in a variety of ways. An analysis of how the Norman and Plantagenet monarchs died provides food for somber thought. William the Conqueror died of a wound received in a siege, already an old man by eleventh-century standards. William Rufus was killed in a mysterious hunting accident. Henry I died of what we would call gastroenteritis, a common fate of monarchs forced to travel in a mobile camp—the same fate overtook King John. Henry II, prematurely aged by endless work and campaigning, died of exhaustion, still soldiering. So did Edward I, an old man in a war camp, his sword by his bedside. Richard I was killed in a siege. Henry III and Edward III were lucky and met peaceful ends after long lives. Edward IV died exhausted; he was also prematurely aged. The same could be said of Henry V's father, Henry IV. Edward II was murdered by having a red-hot poker thrust up his anus. Richard II, Henry VI and Edward V were murdered too, and Richard III was killed in battle and his body stripped naked. It is a daunting necrology, is it not?

Henry V was inducted into this life of toil, risk and exposure in what might be called the hard way. Once, when I was giving a history lesson to the late Princess Diana, wife of the Prince of Wales, we discussed the predicament of a person born to be king. Her husband had been heir apparent from birth. She said she had found him utterly selfish and self-centered because he had been spoiled from the cradle on. I pointed out that this was the common fate of heirs apparent, and that they rarely matured into successful monarchs, unless special circumstances made their childhood or youth exceptionally difficult, hazardous, or stressful. We went through a list of all the English monarchs to see if the theory held up—which it did on the whole.

Henry V fits into this analysis. He was, indeed, heir apparent from the age of twelve, when his father seized the throne from King Richard II, until his twentieth year, when he succeeded as king. But his childhood, youth and early manhood were unsettled, not to say vertiginous, never easy, often difficult and even perilous. His grandfather John of Gaunt, by marital prudence, had become perhaps the richest man in Europe, and Henry's father (known variously as Derby, Hereford and Bolingbroke), as the sole surviving son, inherited it all, and thus became an object of suspicion to the king. In 1398, when Henry was eleven, Richard II exiled Henry's father, confiscated all his property and took the boy as a kind of privileged hostage on his disastrous campaign in Ireland. Had Richard been a more decisive and ruthless man, the boy Henry's life might well have been forfeit when his father invaded England to seek redress.

As it was, by the age of twelve, Henry (often referred to as Henry of Monmouth, where he was born) was heir apparent, and bore the sword of justice at his father's coronation. He performed his first political act: he gave his assent, in secret and along with other magnates, to the strict imprisonment of the former king. He was declared the "Prince of Wales, duke of Aquitaine, Lancaster and Cornwall, Earl of Chester, and heir apparent to the kingdom of England." These were not empty titles. The revolt of Owen Glendower in Wales, trouble in Scotland and the tendency in England of discontented nobles to seek the return of Richard to the throne meant that the king had to unload some responsibility on his young heir's shoulders, and in effect the prince was given charge of Wales. And, since the deposed Richard was too dangerous to the king to be permitted to live, young Henry, barely thirteen, was made party to the decision to have Richard put to death in Pontefract Castle. In discharging his Welsh duties he had, of course, a council, but it is clear he was active both in decision taking and fighting almost from the start. He was present and fought in a siege while still thirteen, and as he moved through his early teens more and more real power shifted to him. At the battle of Shrewsbury, culmination of the desperate crisis of Henry IV's reign, when the Percys of Northumberland joined the other rebels, Henry was aged fifteen and present throughout the battle, receiving a nasty wound in the face but continuing to fight until the death of Percy ("Hotspur") led to a rebel rout. Young Henry was present too in the aftermath, when Percy's uncle, the Earl of Worcester, was publicly executed for treason, and he was given all Worcester's silver.

Henry's teenage training, indeed, was as much in finance as in fighting, for most of the money required to keep an army in the field against Glendower, and to garrison the network of expensive Welsh castles—Caernarvon, Conway, Harlech, and so on—which were vital to holding Wales down at all, had to come from Henry's revenues as Duke of Cornwall and Earl of Chester. Some support he got from the treasury in London, but this had to be negotiated, and Henry had to use his personal clout to ensure that the money, in trunks of silver and barrels of copper pennies, actually arrived at his castles and headquarters where the soldiers were paid. It was hard, anxious work, and good training. Henry was never in any doubt, from his earliest youth, of the close connection between cash and successful military force.

There was also hard fighting in Wales, the prince being in the thick of it, reporting growing success against the Welsh rebels in letters to his father, and being repeatedly thanked for his efforts both by Henry IV and Parliament, which pronounced him a youngster of "bone coer et corage." By the time he was eighteen, the Welsh uprising had been mastered and the speaker of Parliament probably gave him the credit. By this time he was attending Parliament himself and is also listed as sitting in the King's Council. He was a grown man, and an experienced one with five or six years of varied campaigning behind him, including two successive sieges which he conducted personally. The documents show him dealing with artillery experts, "canoners et autres artificies," cannon stones, saltpeter and gunpowder. The first great age of gunpowder was dawning, and Henry was right in the forefront of the new technology. The evidence suggests that it was the prince's use of cannon which proved the decisive factor in his suppression of the Glendower revolt, and also his ingenuity in employing ships to move the guns around the royal castles to Wales, most of which had direct access to the sea. By being a professional soldier very young, and acquiring the skill and jargon of the trade, Henry also formed working friendships with those members of the nobility who took fighting seriously, and the knights who formed the officer backbone of the royal army. These two powerful groups in society, many of whom later served him in France, became loyal subordinates and treated him with the highest respect. It cannot be overemphasized that the extent to which a king knew the business of commanding soldiers in battle, and his personal courage in action, was absolutely vital in ruling a medieval kingdom successfully. Right from the beginning of his career, Henry shaped up as a practical, down-to-earth hero, of the camp and palisade, the siege engine and the battery, the cavalry charge in full armor, and the hand-to-hand combat. He had scars to show too.

Hence the Prince Hal Shakespeare portrayed in *Henry IV* parts one and two—roistering, consorting with low companions, robbing travelers, in trouble with the lord chief justice—has little basis in reality. What is true is that, from 1409, when Henry IV made his will, he was ill and unable to discharge all the duties of kingship. Prince Henry had to take on many of them himself. There were inevitable disagreements over policy, and the prince may have become frustrated. A later Burgundian chronicler, Monstrelet, on what authority we know not, told a story that the prince, attending the sick king in his bedchamber and thinking him asleep, took up the crown, which lay on a cushion by his bedside. At that moment the king awoke. This tale was taken up by the Tudor history writers Holinshed and Hall, and eagerly seized upon by Shakespeare, who had a marvelous instinct for a dramatic scene. In a sense he wrote a whole play around this incident. There was also a tale that the prince came to the help of a trusty servant of his who was being tried for an affray, and the chief justice, William Gascoigne, cited him for contempt of court. There may be some truth to this, though tales that the prince kept company with men of lower rank clearly refer to his friendships with soldier-knights, who had served with him before and were to do so again. The prince's recorded attendance in Parliament, and in the council, his trips to Wales and to Calais, where he was captain of the castle and port, and to his other widespread possessions and responsibilities left him no time for regular dissipation. Shakespeare seized brilliantly on a few fragments to construct a superb play about a debauched young man who redeems himself and matures into a hero. But to my mind the part of Prince Hal, shown as frivolously deceitful and ultimately mean and cruel in his treatment of Falstaff, never quite rings true. And the reason is that it was untrue. The young Henry was a hardworking, serious, and extremely efficient young man who was all along preparing himself systematically to become a great warrior-king.

Where Shakespeare gets the man absolutely right is in his play of Henry as king. The moment Henry came to the throne, in 1413, he began preparations for an invasion of France to assert his territorial claims to Aquitaine and Normandy. He planned not just to conduct cavalry raids, or *chevauchées*, through France, as Edward III and the Black Prince had done, but to take fortified towns and conduct a conquest of firmly held territory. Under cover of diplomacy with the weak French king Charles VI, he stockpiled arms, assembled a siege train and readied ships to transport 10,000 men across the Channel. He confirmed his links with the fighting peers of the English nobility, got Parliament, when he was already liked and trusted, to vote for financial supplies, and sailed from

Southampton in August 1415, landing on the Normandy coast near Honfleur, which he promptly besieged. At once his professionalism asserted itself. He set up his big guns in front of the main gate of what was supposed to be an impregnable walled city. Their bombardment destroyed the gate completely, and the garrison promptly surrendered. He then "stiffened the town with Englishmen" and prepared to turn it into another Calais, a fortified harbor as base for the reconquest of Normandy and a jumping-off point for the southwest too.

Then a typical campaign disaster struck in the form of an outbreak of fever. It carried off some of Henry's closest friends among the army elite, and a third of his 10,000 men had to be sent home sick. His council advised him to return home too, but Henry felt that his taking of Honfleur, although it was "the key to the sea of all Normandy," was not sufficient to justify a large-scale and costly expedition. So Henry, against expert advice, determined to march with what was left of his army across northwest France to Calais, and "see those lands whereof he ought to be lord."

Thus he set out, with 6,000 men, mostly archers, to cover 150 miles through every territory. This was a very risky venture, for a French army of 14,000 was now shadowing him. It suggests that Henry had an adventurous view of strategy, and great confidence in his own skill at tactics and trust in his own men not to let him down. The French blocked his chosen crossing of the Somme, and he was forced to move inland to an unguarded ford near Nestlé. By this time the main French army had moved ahead of him, crossing the Somme at Abbeville, and blocked his route to Calais. Henry had no alternative but to fight, with the odds against him more than three to one, and most of the French force composed of steel-clad men-at-arms. Before the battle there was a notable exchange, recorded by the king's contemporary biographer, with one of his senior commanders, Sir Walter Hungerford: "We need 10,000 men to fight such a battle." Henry: "I would not have a single man more than I do, for these I have here with me are God's people." This exchange was worked by Shakespeare into a main theme of the play, "We few, we happy few." And no doubt Henry knew that a general who was also a king, in command of a comparatively small army but thoroughly under his control, had a good chance of bringing off a tactical success, provided he knew what he was doing.

This, indeed, is the key to Agincourt, one of the great battles of history. The position Henry adopted was a one-thousand-yard front behind a muddy field, with thick woods on either side protected by archers, who fortified their positions with anticavalry stakes driven into the ground. When the enormous French cavalry force hesitated to charge into the mud, Henry ordered his archers to advance within bow shot—250 yards—to drive in fresh spiked stakes, and then open fire. Carrying out this maneuver successfully suggests to me first-class training in battle conditions. It worked, for the French, suddenly exposed to lethal arrow fire as they stood, followed their instincts and charged. Then, seeing the stakes, and with the fire of arrows intensifying, they tried to wheel their horses and back off, so turning the ranks of horsemen behind them into total confusion. The French horses were on the whole not big enough to allow a heavily armored fighting man to ride them skillfully in a mêlée. Henry knew this, and so did his bloodthirsty archers. So the great struggling mass of the French army became an easy target for endless volleys of arrows. The horses were easier to kill or wound than their riders, but once they were unhorsed, the archers ran up and clubbed them to death, or slipped knives into the slits in their armor.

The result was a fearful slaughter. Many French knights, including notables, surrendered. But one-third of the French army, unable to get into the conflict, was still intact, and Henry ordered that partly disarmed prisoners who could not be easily controlled should be dispatched. This was done, as was characteristic of Henry's decisiveness and indeed ruthlessness. The scale of the slaughter has doubtless been exaggerated, for the king was not anxious to kill valuable captives who could be ransomed. In any case, we know that 1,000 high-value prisoners were brought back to England. These included some of the grandest and richest nobles of France. The victory revealed Henry as a matchless field commander—confident, cool, quick thinking and thoroughly professional. It was an overwhelming victory, against heavy odds, and established him at a stroke as the greatest general in Europe.

It also had a psychological effect on Henry himself. Already a pious and God-fearing man, he became convinced that his victory was God given, and that he was in truth the Lord's anointed. His father had always harbored doubts about his right to the throne of England, and this had made him a nervous and indecisive sovereign, much troubled by rebellion. Henry's self-confidence became absolute, and it radiated outward—to his own nobles, to Parliament, to the nation as a whole. No one now doubted his title to the throne. His confidence in his destiny, however, was accompanied by what can only be called a humble acceptance of God's goodness to him in saving him from humiliating defeat at Agincourt and, almost certainly, saving his life. Henry never became arrogant. He never pushed his luck. He remained realistic in his demands and expectations. He built up an excellent working relationship with the nobility, the knights of the shires, the burghers of the towns and, so far as we can see, the peasantry. They provided the taxes when he asked for them, but his requests were reasonable, and he was able to prove to them that the money was sensibly spent. His control of the treasury and exchequer was firm and sure. He developed close and good-tempered relations with the Church, which owned a fifth of England. He stood no nonsense. When his uncle Henry Beaufort, bishop of Winchester and reputedly the richest man in England, accepted a cardinal's hat from the pope without getting Henry's permission, the king's response was prompt, dramatic and severe. The new prince of the Church had to abase himself and paid dearly in hard cash for being allowed to keep his red hat. Nor would Henry tolerate the existence in England of religious houses owned by French mother abbeys, whose monks were French and whose revenues were repatriated to France. These so-called alien priories were nationalized. (An unfortunate precedent, as it turned out, for Henry VIII, 120 years later, used the same precedent to seize all the English monasteries.)

Henry, indeed, played the nationalist card with great skill. He was the first English king to speak English habitually and to write it fluently and idiomatically. His letters are nearly always in English. When campaigning in France, he wrote home regularly letters designed to be read out in Parliament or to the leading citizens of London and other towns, reporting what he was doing. He encouraged the use of English in all government procedures, in the courts and in the House of Commons and the Lords. His nobles also began to write their letters in English. Indeed, it could be said that Henry invented the English letter—and in the next generation, as the survival of the Paston Letters shows, it became an art form, a part of English literature. Henry inherited from his father, and grandfather Gaunt, a patron of Chaucer, a love of books. Like his father and his uncle Beaufort, he had a sizable library of his own, for use as well as for show. He acquired his first books when he was aged eight, when he was given seven volumes of Latin bound together (together with chords for his zither). A list exists of 160 books belonging to him as king, and they included poetry, romances, history, law and devotion; in Latin and in French, as well as English for, like Henry IV, the king was at ease in all these languages. He also owned a batch of 110 volumes he had captured at Meaux, though he was probably outdone by his brother John, Duke of Bedford, who acquired over 800 books in France. He was clearly a bibliophile, and so was another brother, Duke Humphrey of Gloucester, the founder of the Bodleian Library in Oxford and whose handsome bust still dominates the oldest part of the institution, known as "Duke Humphrey." Henry had no time for scholarship, of course, but he clearly saw the English language, which had been given an enormous fillip by the work of Chaucer (doubtless familiar to Henry), as part of the national heritage, to be given every support by a patriotic king. There is a tradition that Henry attended the Queen's College at Oxford in 1398, under the supervision of his uncle Henry, who was then chancellor of the university. Indeed, when I was an undergraduate, the part of the college facing St. Edmund's Hall was pointed out as having once contained his sleeping chamber. He was said to have been tallish, well built, athletic, with brown eyes and hair, kept close clipped around the back and sides (like any well-turned-out soldier), clean shaven and with a powerful speaking voice. He was extremely decisive and could be fearsome. Anyone who needs evidence of this last trait should see his handwritten letter, in English, conveyed secretly to the emperor, by hand of John Tiptoft in January 1417, and now in the British Museum Cottonian Manuscripts. The detailed personal ordinances he drew up for his duchy of Lancaster convey the same impression.

Henry entered France again in 1417, this time taking Rouen by siege, making it capital of Normandy again and establishing his authority all over the vast duchy. By a complicated series of negotiations with the Burgundians and the French court, and by impressive displays of military power when necessary—he took every town he besieged, without exception—he entered Paris in 1420, and signed a treaty at Troyes. Under this, Charles VI disinherited his (supposed) son, the dauphin, and made Henry his heir; and Henry in turn married Charles's daughter, Catherine. Henry attended the *lit de justice* at which this arrangement was made, sitting alongside Charles on the bench of state. When Charles died, Henry became king of France as well as England, and was in a military position to enforce his claim to the full. Indeed, most of the French nobility wanted him as king. The dauphin was tainted by the adulterous conduct of his mother and by his reputed participation in an atrocious murder. Henry was known for square dealing with his own nobility and Parliament and above all as a celebrated commander in battle, who could not be beaten. And if Henry had lived another thirty years, who knows how Anglo-French history might have developed? As it was, in 1421 he caught one of the infections which were liable to strike at any moment a man who spent much of his life in unsanitary camps and bivouacs. His death followed the next year. He was in his midthirties, probably the ablest man ever to sit on the English throne, and one who was well liked, respected and trusted by all those, high and low, who had dealings with him. He was generous, too, and thoughtful. He gave his old nurse at Monmouth, Joanna Waring, an annuity of £20 a year—a considerable sum—and his will testifies to his judgment and decency. If I had to pick an unsullied hero from all English history, Henry would be the man.

Many, including some French people, would say that Joan of Arc (Jeanne d'Arc) was France's greatest heroine. Their lives overlapped, just when she was born, at Domremy on the borders of Lorraine and Champagne in eastern France, on January 6, 1412. Henry was twenty-four and soon to become king. When Agincourt was fought she was three. She was a girl of ten, already working on her parents' farm, when Henry died. It is not inconceivable that they might have married. What sons!—what daughters!—might not they have produced! What they had in common was an overwhelming sense of linkage between warfare and the will of God. Henry believed that he was the rightful ruler of France, and that God's support for his claim was demonstrated by his amazing victory at Agincourt and the events leading up to the Treaty of Troyes. Joan believed it was God's will that this entire process should be reversed and that she was God's chosen instrument to place the dauphin, the rightful sovereign, on the throne as Charles VII. Both saw warfare as somehow sanctified, though neither was in the smallest doubt about its uglier side. Both took to it as teenagers, with extraordinary aptitude.

Joan's career was pitifully short. She first saw a light and began to hear voices in 1424, when she was just thirteen. It took her four years to persuade various levels of Gallic authority that she was innocently sincere, and to get access to the dauphin at Chinon on February 23, 1429. He authorized her to put her visions to the test—she put on men's clothes, donned white armor, and collected a force to relieve the city of Orleans, besieged by the English. On April 29, she got into the city, and on May 8, the English raised the siege. There followed other successes, and Joan was able to get the dauphin crowned king of France in the sacral cathedral of Reims on July 17. Between spring 1429 and spring 1430 she was involved in almost continual fighting with the English and their Burgundian allies. In May 1430 she was taken prisoner at Compiègne, and sold to the English for 10,000 livres. Tried in Rouen for sorcery and heresy, she confessed and was sentenced to perpetual imprisonment. When she withdrew her confession, she was burned at the stake on May 30, 1431, and her ashes thrown into the Seine. A quarter-century later, after a six-year investigation, she was formally proclaimed innocent of the charges against her by Pope Calixtus III, in 1456. She was beatified in 1909 and canonized in 1920.

In some ways we know more about Joan of Arc than we do about Henry V, for the official records of her trial, and of her rehabilitation process, have survived, and are a mine of curious information. Yet there are key lacunae. It is tragic that no French writer of her age had the sense to write her life, based on material from those who knew her. It was a miserable age of mediocrity. Henry VI of England was a pious but inactive and feeble ruler. Some of his commanders in France were able but none imaginative or outstanding. The French commanders were, if anything, worse. Charles VII was an odious creature. He took little interest in Joan, reluctantly listened to her, gave her meager support, was never present while she fought his battles, never thanked her for having him crowned—something he could not do for himself—made no contribution to her attempts to take Paris, and no effort to release or ransom her after her capture, issued no protest at her condemnation and did nothing to get the verdict reversed. In this desert of talent and virtue, Joan emerges as a brief candle of courage and goodness, soon extinguished.

Joan was not a shepherdess, more a farmhand. She eventually

learned to write her name, Jehanne, but never to read or write. After her death all kinds of rumors circulated about her supposed royal descent, but there is no evidence of anything unusual about her birth. Her family were totally undistinguished. There is no evidence of any strong influence on her childhood. She loved church bells and, when working in the fields, would kneel down, cross herself and pray when she heard them. But so did countless other peasant girls. She always referred to herself as *la pucelle,* an archaic name for "maid," which has survived entirely because of her. She never showed the slightest desire to marry or have children. Indeed she said at her trial: "There are quite enough other women to perform the usual jobs women do." From the age of thirteen, when she first heard her voices urging her to save France, she was quite clear, coherent and undeviating about what she wanted to do: "to give France her rightful king and expel the English." She always stuck to her original story. It never varied in any important detail. She identified the voices as those of St. Michael, St. Catherine and St. Margaret—always these three. There was no romantic embroidery. It was impossible to move her from it by bullying threats or cunning traps. She evidently had the gift of recognizing people she had never seen before, e.g., the dauphin. She sometimes prophesied, on the whole correctly, but never formally. There is no evidence whatever of witchcraft.

Joan was without vanity—she refused to sit for her portrait. No contemporary image of her exists. We know she had a short neck; little bright red marks behind her right ear; short black hair; a dark and sunburned skin—"black and swart"—a body of "great force and power," "finely and well formed"; she was strong, healthy, plain and sturdy. No one described her as pretty, though the Duke of Alençon, who liked and admired her, said she had shapely breasts. To use a phrase of Jane Austen, "She was no more than a fine girl." But she was formidable. Her original attire was the coarse red skirt worn by the farm women of Lorraine up to the First World War. When she was first allowed to take part in the war, she promptly put on male clothes supplied by her cousin Durand Lassois. Then she was bought a complete outfit, boots and all, by the citizens of Vaucouleurs: black doublet, short dark gray tunic, high boots, black cap. She later wore elements of scarlet and green, the colors of the House of Orleans, to distinguish her in battle, and cloth of gold and scarlet lined with fur (in the winter), but her elaborate appearance was strictly military in purpose. She wore two family rings, inscribed with "Jhesus Maria" and with a cross. She wore plain armor, without gilding or coat of arms. She said her sword would be found in the church of St. Catherine de Fierbois, and it was there, and she wore it always.

The Duke of Alençon, whose life she saved in action, gave her a fine horse. He was impressed by her skill at riding, tilting and in action. She must have been a born rider. This impressed the men too. They were also struck by the fact that she could spend six days and nights without removing a single piece of her armor. They admired too her modesty and the skill and discretion with which she performed her natural functions. In the year or more she spent in the field, there was never any attempt at seduction or rape, by men on her own side or, after her capture, by enemies. This was not, as some writers have supposed, because she was plain. Soldiers in need will rape any woman, irrespective of age or appearance. There was an aura about her which made men respect her. Indeed they testified that, when with her, they were without carnal thoughts. There is no suggestion she was unfeminine, however. Her page, Louis de Contes, who held her in the highest regard, testified that he often saw her in tears. There was something distinctly feminine about the small battle-axe she carried. Her voice was womanly. After the battle of Patay, June 8, 1429, she took the head of a badly wounded English soldier on her knees, saw that he confessed and comforted him until he died. It is significant that she seems to have made a surprising number of women friends. There was no demand among the wives of prominent men, French, English or Burgundian, for her prosecution.

She was wounded twice. The first time an arrow entered her body just above her left breast, penetrating six inches. She pulled it out with her own hands, and quickly returned to duty. At the time of her failure to take Paris, she was hit by an arrow in the thigh, the archer shouting: "Paillarde! Ribande!" (not an Englishman, evidently). In custody she was tortured and ill treated but responded stoically. Her confession reflected loneliness, confusion and bewilderment. She recovered her courage and self-confidence, and her faith, and her last days and hours were impressive. At the stake she called loudly and repeatedly on Jesus. There is a reliable report that John Tressant, secretary to Henry VI, exclaimed as she died: "We are lost! We have burned a Saint!"

The French have always in modern times blamed the English for what happened to Joan. General de Gaulle often said: "I can never forgive the English for what they did to Jeanne d'Arc." But the crime, if it was a crime, was a French one. The two judges were Pierre Cauchon, bishop of Beauvais, in whose diocese her supposed offenses had taken place, and the deputy inquisitor of France, Jean Lemaistre, Dominican prior of Rouen. We have a complete list of those who took part in the trial: 1 cardinal, 6 bishops, 32 doctors of theology, 16 bachelors of theology, 7 doctors of medicine, and 103 others. Of these only eight were English, of whom two attended regularly but took no part in the process. The trial was in many ways a travesty since Joan was allowed no counsel and was not permitted to call any witnesses. Her chief enemy was not really the English but the Church. She had made no effort to conciliate the French clergy and as a free woman had operated entirely through the secular authorities. She does not seem to have had much to do with the clergy at any time in her life. What made Joan dangerous, in the eyes of the Church hierarchy, was that she claimed to have a message directly from above, without any mediation by the Church. So did many other antinomians, eccentrics and egregious religious spirits in the fourteenth and fifteenth centuries. The Church invariably treated them with suspicion and hostility, and many were prosecuted for heresy. In the Church's eyes, the fact that Joan was an enemy of the English was secondary, though obviously the Church had to take note of the views of what was theoretically the strongest crown in Christendom. But English official hostility merely reinforced the horror and detestation the Church felt for a young woman who used the power of religion outside the parameters they laid down. It would be accurate to say that Joan was the victim of clerical trades unionism and male prejudice.

The heroic glamour of Henry V and the tragic faith of Joan gave a fine glow of nobility to the last phase of the Hundred Years War, otherwise a soulless and wasteful episode in European history, which reflected no credit on England or France, or the universal church for that matter. And here it is worth remarking that, whereas the English, through the pen of Shakespeare, made Henry V the exemplary poetic hero of a splendid play, the French failed to turn Joan, who ought to have become a transcendental national heroine, into an artistic paradigm. There was a failure of kingship to begin with. Charles VII did nothing for Joan, though he owed his throne to her. The delay of more than a quarter century in the rehabilitation procedure was significant. And the verdict was shifty. It did not declare Joan a martyr or even actually state that she "had remained faithful and Catholic until and including her death"—as her family had wished—but merely that the judges had "acted improperly." None of the key judges, all French, was condemned. Some were still alive, but they were left undisturbed in their benefices. The French kings exercised enormous influence in Rome, where they were always addressed as "Your Most Christian Majesty." But none pressed for Joan's canonization. It was left for the Third Republic to do that, and to declare her "the second patron of France," with the second Sunday in May as a national feast in her honor. But her day has never been a public holiday.

French writers have been curiously reluctant to use Joan as a theme. It is true that Jean Chapelain (1595–1674) wrote an epic, *La Pucelle,* in twenty-four cantos. But though he was a great friend of Richelieu, only half was published—the rest had to wait until 1882. Only a handful of French have even read it; the vast majority have never heard of it. It is true that Voltaire wrote *La Pucelle*. But it is not one of his better works. There is a famous portrait of Joan in Jules Michelet's history of France (volume five), a prose life by Anatole France, a poem by Péguy, and a play by Paul Claudel, *Jeanne au bûcher,* set to music by Honegger. Anouilh also wrote a play, *L'Alouette*. But there is nothing in French literature comparable to George Bernard Shaw's masterpiece, *Saint Joan*. And a number of English churches have been dedicated to her. Someone had the justice and wit to erect a statue of her in Winchester Cathedral, opposite the tomb of Cardinal Beaufort, who had a discreditable, albeit marginal, role in her condemnation. When, after the appalling scenes of murderous atheism that marked the Paris commune of 1870–1871, the great basilica of reparation was erected on the summit of Montmartre, a matchless opportunity was missed to dedicate it to Joan. But of course she was not then a saint! The truth is, from start to finish, from the miserable ingratitude of Charles VII onward, official France has always shown itself unworthy of this great patriotic martyr. But she dwells warmly in the hearts of many people in France, even in this infidel age, just as Henry V and his happy few are remembered by those English who still love their country.

5

HEROISM IN THE AGE OF THE AXE: ST. THOMAS MORE, LADY JANE GREY, MARY QUEEN OF SCOTS, ELIZABETH I AND SIR WALTER RALEGH

The Tower of London is an unrivaled breeder and decapitator of heroes. It bestrides the Roman wall and there has been a fortress on the site since the late first century. William the Conqueror built the White Tower here, from 1077, of Caen stone, whitewashed. Thereafter it contained royal apartments, a chapel, an armory, a mint, a zoo (mainly families of lions and an occasional leopard), barracks and ample dungeons for state prisoners. During my time in the army, it was still a regimental headquarters, of the Royal Fusileers, and I remember dining within its grim walls with the officers. I defy anyone, even today, to visit this place without a shiver of fear. The graffiti of desperate, lonely, scared and brave prisoners are still on its walls, some half a millennium old.

The tower was carefully graded to distinguish both the ranks of prisoners and the degree of animosity the monarch felt toward them. King John II of France, captured by the Black Prince at the Battle of Poitiers in 1356, and Charles, Duke of Orleans, a captive from Agincourt, both occupied the royal suite normally used by the English sovereign the night before the coronation. But these were ransomable prisoners, who had to be kept alive in order to retain their cash value. Some high-ranking prisoners got poor accommodation. Much writing, including poetry, was done by prisoners. Sir Walter Ralegh wrote the entire first part of his *History of the World* there, in a large, light-filled room. Those of princely rank and of other nobility were never tortured, gentlemen rarely. The most merciful death was decapitation by a giant sword: the executioner at Calais had to be brought over specially to do this, a privilege accorded usually to great ladies. It was also regarded as a privilege to be beheaded privately on Tower Green, rather than publicly on Tower Hill, outside the walls, in front of a bloodthirsty and jeering mob. Lower-class victims or those the monarch really hated were hanged. But this too had degrees of cruelty. Orders could be given that the man (or woman) have the neck broken the instant of pendulation. This was merciful. But, strictly speaking, the judge ("guided by the monarch") in sentencing did not need to order "that ye be hanged by the neck until ye are dead." The sentence of "hanging, drawing and quartering" implied that the condemned person be taken down "while still quick" (i.e., alive), then have his entrails drawn, and burned in front of his nose while he could still smell them. As a further disgrace, the body was cut into four quarters, and each displayed in a particular public place until decayed. A victim was saved from the worst extremities only if he or she cooperated with the authorities, at the trial by confessing and pleading guilty, and on the scaffold by telling the crowd that the sentence was just and asking God to bless the sovereign. Henry VIII, who executed more people than any other English monarch, was particularly insistent on these points. In short, the sentence of death was only the beginning: the agony of your end was determined by your humility and public self-abnegation, or your determination to have the last word.

The last celebrity executed at the tower in public was Lord Lovat, hanged for his part in the 1745 rebellion of Bonnie Prince

Charlie. Lovat, aged eighty-two, kept alive the tradition that a great man died with spirit. On his way to the scaffold, a hag screamed out: "They're going to hang ye, ye old Scotch dog," to which he replied: "I believe they will, ye old English bitch." But the tower's age of heroism was the sixteenth century, when a cast of exceptional characters played their parts on its grim stage.

Noblest of them all, in life and in death, was Sir Thomas More, former lord chancellor, executed by axe on July 6, 1535, for refusing to acknowledge Henry VIII as head of the church, as decreed by Parliamentary statute. More is one of the earliest historical characters who comes to us as a real, living, rounded personality, thanks to the huge increase in documentation which began in the early sixteenth century, and the Renaissance cult of individuality which is reflected in memoirs. Not only his sayings but his tone of voice seems to survive. And we know exactly what he looked like, for Hans Holbein the Younger painted him among his family, gloriously and intimately, and in a superb half-length, now in New York's Frick, in which More's face, both formidable in intellect and resolve, but fragile in its capacity for feeling and suffering, positively glows with life; who would not want to possess this marvelous image of the finest spirit of an age which is here reanimated for us, to instruct, warn and move?

More has a lot to teach the twenty-first century. He had a curiously modern gift for words, and he articulated the culminating drama of his life—the still, small voice of conscience defying an ideological despotism—with stunning aptness, so that to us he resembles the hero of a contemporary morality play. He could be compared to George Orwell, though it must be added that the ideal society described in More's *Utopia* is not entirely unlike Big Brother's state in *1984*. There, people guilty of privately discussing matters of public interest may be put to death: it is also the only crime in Utopia (apart from adultery) that is punished capitally. In Utopia, the citizens are expected to check up on each other all the time. More was an anti-individualist in many ways. He trusted the group as opposed to the independent-minded man. So why did he pit his own conscience against most of the English bishops? Because, he argued, the rest of Europe was on his side: he would not have trusted his own conscience alone. More is full of such paradoxes.

The key to More's thinking, to what made him a hero and a martyr, can be found in his unfinished biography of Richard III. More's portrayal of Richard as a ruthless usurper is often dismissed as a work of Tudor propaganda written to curry favor with the regime, itself made possible only by Richard's overthrow. But this interpretation is wrong. For too many powerful men and women who had a direct interest in that drama were still alive in 1514–1515, the years during which More was working on the biography. Far from advancing his career, publication would have destroyed it. It did not see print till 1557, long after More's death. Then, it is true, it became the model for the treatment of the chroniclers Thomas Hall and Raphael Holinshed of Richard's tyranny, and so of Shakespeare's stunning play, and the standard version of the reign. But it expresses not so much Tudor propaganda as More's political philosophy.

In the person of Richard III More presents the perversion of a divine institution. Kingship, instituted by God as a source of authority, could be turned into tyranny, just as the church could be corrupted by the love of money. More uses the character of Thomas, Cardinal Morton, in whose household he had served and with whom he must have discussed Richard, to describe the wrong response to tyranny. Morton argued that Richard's usurpation must have been God's will because it had happened, and therefore it was the subject's duty to accept it. Morton was an early *politique,* articulating a de facto theory of government that had been slowly emerging during the turbulent dynastic revolution of the Wars of the Roses. Later the theory was to take parliamentary form, so that in the seventeenth and eighteenth centuries whomever Parliament pronounced king by statute *was* king, and sovereignty resided in "the king in parliament."

More rejected this pragmatism not only because it was contrary to natural law, not only because it was a form of moral relativism as opposed to the absolute morality upheld by the universal church, but also because it reflected the doctrine of predestination associated with the reformers—Luther, Tyndale and, later, Calvin. More not only accepted but passionately supported authority in the legitimate discharge of its natural-law functions, and just as passionately opposed it when it assumed the caricatured form of a tyranny. The study of Richard III thus became a theoretical preparation, a dress rehearsal for his own conflict with Henry VIII. In More's view, a legitimate king, who should be a shepherd to his flock, could become a tyrant and a devouring wolf, imperiling not only their bodies but their immortal souls. Here was another paradox. More would have had no difficulty in facing up to the modern notion of a "parliamentary despotism" or a "presidential tyranny," a juggernaut state destroying its own legitimacy by expanding beyond its ordained role. The solution lay in a separation of powers: an ideal Christendom, for example, had the universal papacy and the individual monarchies balancing each other. He would have approved of the separation of powers in the American constitution, though favoring a constitutional document which was itself based in natural law. Henry VIII's statutes defying the papacy began with the phrase "This Realm is an empire." In More's view no realm was, or could be, an empire—that is, a state acknowledging no superior power. In defying Henry VIII he was resisting both a false idea of kingship and a false idea of the state.

More was thus a hero of the highest category because the stand he took, knowingly paying the price with his life, was based upon universal principles of absolute morality, applicable in all places and ages. And his sacrifice of life to a principle was conducted on his part with extraordinary dignity throughout, almost with a kind of elegance. The behavior of Henry VIII and his pliant instruments was, by contrast, petty and despicable. Henry wanted to break More's will, and imprisoned him in the tower for over a year before killing him. The accommodation was poor for a man who had been third in precedence in the realm, after Henry himself and the archbishop of Canterbury. He set upon More, to trap him, a mean and shifty creature called Sir Richard Rich, a liar and perjurer for whom More had nothing but contempt. One of Henry's tactics was to remove the books that More had originally been allowed to take to prison. First some went, then all. Rich arrived with two men, who bound the last books up and put them in a sack. After they had gone, More pulled the blinds down and thereafter sat in the dark. When the jailer asked him to explain, More said: "Now that the goods and the implements are taken away, the shop must be closed." Despite his depressed spirits, More responded to Rich's provocative questions with skill, for he was a very resourceful lawyer and dialectician, and his defense of himself, when finally put on his trial, as recorded by his son-in-law, was manly and powerful. His eighteen judges were an unimpressive collection of men who had long surrendered their judgments to Henry's will. They included Thomas Cromwell, at present his instrument, soon in turn his victim; Anne Boleyn's father and brother, the Duke of Norfolk, whose niece Katherine Henry would marry and murder in due course; and Charles Brandon, Duke of Suffolk, who was Henry's closest crony and who had married his sister. Unable to shave, More had grown a long, gray and tangled beard, and he and his clothes were dirty as a result of his close confinement. But his voice was clear and firm. His eloquence and legal artistry was wasted, of course, on his judges, who were in a hurry to pronounce verdict and sentence. But they are not wasted on us.

Bishop Fisher of Rochester, More's fellow accused and companion in suffering, had already been executed. He was condemned to be hanged, drawn and quartered at Tyburn. The Carthusian priors who had resisted Henry's will had earlier been dispatched there, to inspire terror in Fisher and More. Though hanged, they were still conscious when their bowels were taken out and burned, and one account says they still breathed and felt when the hangman cut into their chests to take out their hearts. Fisher, in the end, was so sick his sentence was reduced to beheading on Tower Hill. He looked frail and half-starved, "like a skeleton," but he showed stoicism and good humor. A hero needs to know how to die. Fisher managed to get up the rickety scaffold without help, and spoke to the crowd: "Hitherto I have not feared death. Yet I know that I am flesh and that even St. Peter, from fear of death, three times denied the Lord. Wherefore help me with your prayers, that at the very instant of my death's stroke, I faint not in any point of the Catholic faith for any fear." One of the advantages of a public execution was that the crowd could offer up their prayers for the victim, and be seen—by him and others—to do so.

All this was reported back to More. He did not know until shortly before his own execution whether or not he would be spared hanging and the horrors that followed its half completion. In the end Henry did not risk the revulsion that would have followed treating his old lord chancellor like a common murderer, and allowed him the mercy of the axe. Dame Alice, his wife, had cursed him for his lack of worldliness in disobeying the king, but his daughter Margaret (married to Roper, his biographer) was his staunch supporter to the end. The day before his death, More wrote his last letter to her, sending it with the hairshirt he had always worn:

> Our Lord bless you good daughter and your good husband and your little boy and all yours and all my children and all my godchildren and all our friends. Recommend me when you may to my good daughter Cicily whom I beseech our Lord to comfort, and I send her my blessing and to all her children and pray her to pray for me. I send her a hankerchief and God comfort my good son her husband. I cumber you my good Margaret much, but I would be sorry if it should be any longer than tomorrow, for it is St Thomas Eve and the Utas of St Peter, and therefore

> tomorrow long I to go to God; it were a day very meet and convenient for me. I never liked your manner towards me better than when you kissed me last for I love when daughterly love and dear charity hath no leisure to look for worldly courtesy. Farewell my dear child and pray for me, and I shall for you and all your friends that we may merrily meet in heaven. I thank you for your great costs. I pray you at time convenient recommend me to my good son John More. I liked well his natural fashion. Our Lord bless him and his good wife my loving daughter to whom I pray him be good, and he hath great cause.

More was beheaded on the morning of July 6, 1535. Sir Thomas Pope, who brought him the news, and who was a friend, told him that the king's pleasure was that "at your execution you shall not use many words." So More did not prepare or make a speech from the scaffold. He planned to die in his best clothes, but at the insistence of the lieutenant of the tower, he "dressed down." His clothes at death became the property of the executioner, by tradition, and the lieutenant said the man was a rogue and would sell the garments and spend the money riotously. So More borrowed an old gray cloak and put it on. But he gave the rogue an angel (13s. 8d.) all the same. He did not like the look of the scaffold and said: "I pray you, Master Lieutenant, see me safe up, and for my coming down, let me shift for myself." More embraced the executioner, kissed him, said he forgave him and gave him his blessing. He said to the people: "Bear witness with me that I shall now suffer death in and for the faith of the Holy Catholic Church. I live the King's good servant, but God's first." To the executioner: "Do not be afraid to do your work. My neck is short, and if you care for your reputation, do not strike awry." He then bound his own eyes with a linen band. The block being low, More lay on his stomach to put his head across it. He moved his beard out of the way, "as it has done no treason." Those were his last words: an ironic joke. Next to the future Jane Austen, More dealt in irony more often than any other character in English history. But there was nothing ironic about his heroism.

The Duke of Suffolk, one of More's judges, and a coarse but proudful man, did not then know that his own granddaughter, as yet unborn—she came into the world two years later—was to meet the same fate as More in that lethal process of religious conflict and dynastic ambition which kept the Tudor scaffold in the tower so busy. Lady Jane Grey (1537–1554) was the daughter of Frances Brandon, child of the duke by his marriage to Henry VIII's sister, and so great-niece of the monster. Highborn ladies suffered and feared in those days because monarchs bred so few sons. Her short life—she was not yet seventeen when she was beheaded—was both miserable and elevated, even glorious. Miserable because, all her days, she was the powerless victim of the cruelty, ambition and ruthlessness of her elders. So far as we can tell, she was herself a good, obedient, sensitive and high-minded child, both as infant and teenager. All those who dominated her life were worthless people. Her parents, Henry Grey, Marquess of Dorset (later Duke of Suffolk) and Frances Brandon, brought her up with great severity, "more than needed for so sweet a temper," according to a tradition reported by Fuller in his *Worthies*. She herself said she was constantly punished "with pinches, nips and bobs." At nine, she entered the household of Henry VIII's last wife, later widow, Catherine Parr, and so came under the shadow of her evil second husband, Lord Thomas Seymour, who purchased her wardship when Catherine Parr died. During the protectorate of the Duke of Somerset, there were rival plans to marry her to the young King Edward VI and to Somerset's heir, the Earl of Hertford. They were frustrated by Thomas Seymour's execution, then the beheading of his elder brother, Protector Somerset. Thereafter Jane fell into the hands of the rival clan, the Dudleys, under the protectorate of the head of the clan, the Duke of Northumberland. Jane was married on May 21, 1553, aged fifteen, to Guilford Dudley, Northumberland's fourth son. The Venetian ambassador reported that Jane had strenuously resisted the marriage, but was forced to yield by her father's violence. After her marriage she was forced to live not only with her husband, whom she viewed with fear and distaste, but with her parents-in-law, the Northumberlands, whom she came to hate. Her distress led to a serious illness, from which she nearly died.

She lived thanks to the other element in her life: learning. Women in the sixteenth century had to put up with dreadful oppression. But they were often given the opportunity to acquire an excellent education, and to acquire a delight in the world of books and philosophy denied to them in life. What amazes us today is the relish with which the clever ones took to scholarship. Jane's parents beat her because she did not take to prideful deportment or sewing, the only accomplishment they thought fit for a girl destined for a grand marriage. But Jane was fortunate in being given the Reverend John Aylmer, afterward bishop of London, as tutor. His delight in the eagerness and brilliance with which she learned Latin, Greek and Hebrew was exceeded only by her own glee and satisfaction. She was a beauty too, and the sight of her fine head bent over her books almost broke Aylmer's heart. In the summer of 1550, when she was twelve, she was visited by an even greater scholar, Roger Ascham, who found her reading Plato's *Phaedo,* while the rest of her family were busy hunting deer in the park. That December, Ascham wrote a letter to his friend Sturm saying that her skill in writing and speaking Greek was almost beyond belief. She learned French and Italian too, as a matter of course, and wrote letters in all these languages to famous scholars on the Continent, who were friends of her tutors.

Her marriage to Dudley, who was barely literate, was an intellectual penance as well as a physical one, and perhaps a social one too, since Edmund Dudley, founder of the family's fortune, was an upstart, a finance minister and extortionary, created by Henry VII from nothing and hanged by Henry VIII on his accession, to please the public. Worse came, for when Edward IV died of tuberculosis on July 6, 1554, opening the road to a Catholic restoration under Henry VIII's Catholic elder daughter, Mary Tudor, Northumberland, to prolong his power, brought Jane and her husband, his son, to the tower, and had them proclaimed queen and king. Jane signed, or was made to sign, various documents "Jane the Quene," but to what extent she acquiesced in her usurpation we do not know. Being educated, indeed learned, for a sixteen-year-old, she must have known it could not possibly succeed. She had been brought up "reformed" but cannot have believed it was God's will that the succession should be wrenched from the rightful claimant, Mary. She is recorded as saying that she loathed and repudiated her father-in-law and his schemes. As she expected, the plot to make her sovereign failed ignominiously. Mary was everywhere greeted as the rightful sovereign, and after eleven days, Jane passed from being a queen in the tower to being a prisoner. Four months later she was charged with high treason at the London Guildhall, appearing in "a black gown of cloth, a French hood, all black, a black velvet book hanging before her, and another book in her hand, open." She pleaded guilty and was sentenced to death. The hopeless rebellion of Wyatt, in which her worthless father took part, made it inevitable that the sentence was carried out. She and her husband were both beheaded the same day, February 12, 1554, she witnessing his death just before her own. She said a few words from the scaffold, saying she had never wanted the crown and that she died "a true Christian woman."

Was she a heroine? We know pitifully little about her or her thoughts. The few letters from her that survive are moving but uninformative. We do not know for sure what she looked like since all the supposed portraits of her—new ones are added from time to time—are matters of dispute. A piece of her embroidery was kept for centuries but is now lost. She was long revered in the minds of well-brought-up Protestant English schoolgirls. Nancy Mitford told me that the only way she had ever been able to obtain a satisfactory orgasm while masturbating was by thinking about Lady Jane Grey, which produced an extraordinary frisson of ecstatic fear. But it is not clear that teenage girls today know anything about her.

That cannot be said of Mary Queen of Scots. British publishers say that only three biographies can be pretty well guaranteed to sell well: those of Napoleon, Byron and Mary Queen of Scots. She reaches out over the centuries to pluck at heartstrings and arouse passions. She is closely associated in the minds of modern aficionados of history with Queen Elizabeth, as indeed she was among their contemporaries: they were termed then "the Daughters of Debate." No one who regards Mary as a heroine, and many do, will accord the same accolade to Elizabeth. Or vice versa. They make an antagonistic pair, rivals for our sympathy, like Cavaliers or Roundheads, Unionists and Confederates. Mary, in particular, is a classic case of hero worship. She cannot be assessed dispassionately as a political strategist and stateswoman, as Elizabeth can. She has to be loved and admired as a person. And those who favor her have to do so all the more fiercely because she was guilty of undoubted follies and, many would say, crimes.

Mary Stuart was born on December 7, 1542. She was thus nine years younger than Elizabeth Tudor, twenty-six years younger than Elizabeth's Catholic half-sister, Mary Tudor, and five years younger than Lady Jane Grey. All these women had claims to the English throne. Mary Stuart's came from King James IV of Scotland's marriage to Margaret, eldest daughter of Henry VII and Henry VIII's sister. Their progeny was James VI, Mary's father, and this line eventually proved the winner, since her son James became the tenant of both thrones, being James VI of Scotland and James I of England.

James V, Mary's father, was a pathetic creature, saddled by a persistent ill fortune that marked her own life. His brilliant father was slain at the disastrous battle of Flodden in 1513, and he inherited the throne as a small child, under a succession of villainous regents and with the country in a state of anarchy. He opted for the "Old Alliance" with France, and married François I's daughter Magdalene and, on her early death, the heir presumptive of the House of Guise, Mary, whose daughter, named after her, was the sole progeny of the union. James three times attempted to invade England, but his army refused to follow him over the border, and on the last occasion it was scattered by a band of English freebooters at the battle of Solway Moss. James died of a fever contracted in camp, seven days after the birth of his daughter. (Two sons had died in infancy.)

Until his death in 1560, Mary of Guise, who had earlier been married to Louis d'Orleans, and was an adopted daughter of François I—and so sister to his successor Henri II—was deeply involved in French politics but also, intermittently, regent and ruler of Scotland. John Knox, the fierce Protestant ayatollah of Edinburgh, saw her as a Catholic enemy, accused her of rejoicing at the death of her royal husband (he had recently taken a mistress) and of having affairs with Cardinal Beaton, head of the Catholic party in Scotland until his murder in 1546, and with Patrick Hepburn, third Earl of Bothwell. She was certainly a desirable and fascinating woman, who on her occasional visits to France "was almost worshipped by the court," and "bore its whole swing." She was devious, two-faced, full of promises to both sides, shed plenteous tears when suitable and was free with her smiles and caresses. When reminded of her broken promises to the Protestants, she replied tartly that princes must not be tied down to their words, and said she would "banish ministers [for preaching Protestantism] even though they preached as truly as St Paul." But she had an impossible role, balancing Protestant England against Catholic France, and the various factions among the Scots nobility, the Protestant burghers of Edinburgh and other towns, and overcoming the propaganda of John Knox, preaching against "the monstrous regiment of women," and subjecting her personally to a brutal campaign of lies and half-truths. In her last years she suffered from dropsy and lameness, and much of her time was spent defending fortresses or bargaining with "lewd Scots" as she called them. She personally had captured rebels hanged outside her stronghold of Leith, and Knox described her as "hopping with joy" at the sight of "the corpse of the saints," and exclaiming (according to him), "Yonder are the fairest tapestries I ever saw." The rights and wrongs of these dreadful years, or even the truth of the bare events, will never be known—we have to imagine a tartan version of Afghanistan—and death must have come as a merciful release to this spirited but overburdened woman.

Mary of Guise, and her attempts to govern Scotland, are worth describing, because her daughter Mary faced much the same problem, which she tried to deal with in a similar way, being a similar woman. She took after her mother rather than her father, inheriting good looks and a tall, slender, graceful body; high spirits and love of fun; courage, sensuality, and deviousness. Mary was devious from first to last, with little respect for the truth. But then, she never possessed the power to be straightforward, and could seldom afford to speak the truth.

Mary of Guise made certain her daughter would be brought up in the French and Catholic interest by arranging for her to marry François, son of Henri II and dauphin of France. She was shipped to France in 1548, aged six, and educated with other French princes and princesses, under the supervision of Henri II's sister Margaret, a learned lady. Some Latin exercises of hers, aged twelve, have been published, and she learned some Greek and Italian. French was her mother tongue. Some poetry by her survives, as well as spurious stuff, but is unmemorable. We also have examples of her needlework, nothing special. As she grew, all at the French court testified to her beauty. Her height was unusual, five feet eleven. All three of her husbands were shorter than her, two much smaller in fact. She had red-gold hair and amber-colored eyes. The cardinal of Lorraine, one of her uncles, said that he had never seen anything to approach it among the daughters of royal, noble or commoner houses. He also testified to her staunch grasp of Catholic doctrine, and her strict morals. In fact the tone of the French court, though cultured, was meretricious, and when her mother visited her in

1550, she decided to remove her to a more salubrious setting. But nothing was done. Mary was married to François in 1558, and the next year Mary became queen of France on the death of Henri II. But her husband was sickly, and died at the end of 1560, soon after her mother. Mary was then left alone. Two years before, on the death of Mary I of England, Mary Stuart had claimed the English throne, on the grounds that Elizabeth was illegitimate. Elizabeth became queen nonetheless, and in a reign of forty-five years her title was never seriously challenged. Elizabeth was the queen of the Protestant interest, and Mary the putative queen of the Catholic one. She was also the queen of the Catholic interest in Scotland, but that was a diminishing asset and her sovereignty there was never secure. It is true she had the support of France, both in asserting her right to the English throne and in supporting her exercise of it in Scotland. But France was increasingly divided by religious factionalism, close to civil war at times, and was unable to make her true weight felt in Scotland, still less in England. Indeed much of the time France needed English support in her contest with Philip II of Spain, and the Habsburg interest. Mary's position in Scotland was never strong, even at the beginning, and grew steadily weaker. Elizabeth, by contrast, strengthened the regime steadily, year by year. Hence there was never much doubt who would emerge the victor from the contest between the two queens, though the outcome was not so clear at the time as it is to us.

Once François II was dead, his mother, Catherine de' Medici, made it clear that Mary was not wanted in France, and she made haste to Scotland to claim her throne. Things went well at first, primarily because Mary was, as queen regnant of Scotland, and with powerful French links, and a claim to England too, "the best match in her parish." She did her best to impart a little gaiety to the grim Scottish court. Knox complained that, at council meetings, "she kept herself very grave, but as soon as ever her French fillocks, fiddlers, and others of that band gat the house alone, then might be seen skipping not very comely for honest women." Elizabeth, in

London, asked the Scots ambassador if Mary danced better than she did, and was tactfully told that she danced beautifully, "but not as high or disposedly as Your Grace." While waiting for the best suitor to emerge, Mary toured the Catholic north of Scotland, often "living rough" and seeming to enjoy the primitive conditions. The English ambassador, Randolph, wrote:

> In all these garboils I never saw her merrier. She said she regretted she was not a man to know what life it was to lie all night in the fields, or to walk on the causeway with a jack and a knapschulle, Glasgow buckles, and a broadsword.

Among her suitors were the kings of Sweden, Denmark and France, the Archduke of Austria and the reigning Duke of Ferrara, as well as a clutch of French princes and numerous Scottish earls. Mary herself fancied Don Carlos, son and heir of Philip II of Spain, the best choice from a political point of view if she was determined to play the Catholic card and aim at England. But Carlos was "unavailable," and Mary's own choice suddenly (1565) fell on Lord Darnley, son of the Earl of Lennox. He was her cousin, son of a granddaughter of Henry VII, and thus marriage to him would strengthen the English claim of any child born to them. Apart from flashy good looks, however, he had nothing else to recommend him, being without power or wealth. She soon became disgusted by his vile manners, constant debauchery and sheer arrogance. She relied greatly for comfort on her Italian factotum, David Rizzio, to whom she could talk in Italian, which none of the others understood. Darnley aimed at a personal government of his own, second to his heirs, and when Mary resisted, saw Rizzio as his enemy. In March 1566, he and a gang of Scottish Protestant chiefs broke into the queen's cabinet, and Darnley held her screaming, while the others dragged Rizzio into the antechamber and stabbed him to death with their dirks.

This atrocious crime, the direct result of the queen's foolish choice of a husband, resounded through Europe. It marked her as a woman of ill fortune, who was always in trouble. Unfortunately it was only the beginning of her sorrows. A son, the future James I and VI, was born, and Mary appeared to have forgiven Darnley when he fell ill of smallpox. She visited him at a house beside the Kirk o'Field, just to the south of Edinburgh's walls. On February 9, 1567, two hours after Mary had left from a visit to his bedside, the house and Darnley were blown to pieces by gunpowder. The murder was organized by the Earl of Bothwell, and the queen's connivance was assumed, especially as three months after the crime, she married Bothwell, whom she created Duke of Orkney. The marriage was particularly scandalous as Bothwell, to make it possible, divorced his recently espoused wife. Here was a second scandal, worse than the first. This led in rapid succession to a general rebellion of the nobility, the ignominious desertion of her army on the fields of Carberry Hill (June 1567), her imprisonment at Lochleven and abdication in favor of her son, her escape (May 1568) and her final defeat at Langside, near Glasgow. She fled to England and so became Elizabeth's prisoner. (Bothwell ended his days in perpetual solitary confinement in a prison in Denmark, where he had fled; all Mary's husbands were unlucky.)

Elizabeth refused to see Mary, and had her placed under house arrest with various noble custodians such as the Earl of Shrewsbury. There she remained eighteen years while the queen and her counselors argued about what to do with her. She was moved from time to time, since nobody wanted the expense and anxiety of keeping her in custody—she was allowed to hunt and ride for exercise—and so the unfortunate woman was passed on like an unwanted parcel, from Carlisle to Bolton, to Tutbury, Wingfield, Coventry, Chatsworth, Sheffield, Buxton, Chartley and Fotheringay. She knew the ways of every freezing house in the north of England. All her hosts liked her, and found it hard to be strict. "She could charm the stone lions on the balustrade" at Tutbury. Her notoriety was enormous, reinforced by the discovery of letters in a casket belonging to Bothwell, which seemed finally to provide proof of her involvement in Darnley's murder. The casket letters are mostly forgeries but contain, no doubt, genuine elements. Mary's moral character, as generations of historians have found, is a maze, a mystery, an insoluble enigma. She said her prayers, and sincerely too; but she prayed as a sovereign to a maker to whom alone she was answerable. Her belief in her sovereignty was absolute—it was the one constant certitude in the whole of her confused and confusing political career—and in the pursuit of sovereignty Mary judged all devices lawful. If this had been her sole obsession, Mary might have done well. But it was balanced by impulsiveness, a propensity to take sudden, rash and breathtakingly foolish decisions—to marry Darnley, to get rid of him, to run off with Bothwell, to throw herself on Elizabeth's mercy: these were only the outstanding ones. Of course this womanly impulsiveness, on a regal scale, is precisely the trait that makes her so attractive to so many people, then and now. She is real, alive, her eyes flashing, her heart pounding, her body quivering with emotion. We feel it. She performs for us, as in a theater, from the dry documents which record her exploits. For many she contrasts warmingly with the icy calculations of Elizabeth.

That Mary was accessory to murder counted for little among her admirers then (or now). The sixteenth century, the age of religious hot war and cold war—and civil war—was the age of state murder. The fifteenth century, the age of Machiavelli's *The Prince,* had been bad enough; but the atmosphere engendered by violent religious persecution led to a collapse in the already precarious ethical standards of conduct between states, and particularly between princes. Princes broke their word constantly, lied systematically, and used any means whatsoever to ensure their survival. Elizabeth's chief minister, Burghley, believed this was done by divine sanction:

> All forces and resistance begun upon urgent and necessary occasions for the safety of any state whose ruin is greedily sought after, is allowed of God, conformable with nature and disposition of men and by daily experience continued through the whole world.

Machiavelli's book, it is interesting to note, was banned in England as immoral. But it circulated and was often quoted by clerics and politicians alike, Burghley included. He wrote: "I must distinguish between discontent and despair, for it is sufficient to weaken the discontented, but there is no way but to kill desperates." Philip II, "His Most Catholic Majesty," regularly employed murder, judicial or otherwise, as an act of state. He put his own son and heir, Don Carlos, on secret trial, walled him up in his quarters and later announced he was dead. He deliberately and openly plotted the murder of the prince of Orange, William the Silent, put a price on his head, and in 1584 saw his plan carried out. He had two of the greatest members of the Flemish nobility, Counts Hoorne and Egmont, condemned to death and decapitated, in defiance of all tradition and statute. He had Egmont's younger brother garrotted while under safe conduct. He approved and financed plans for the murder of Elizabeth (his sister-in-law, let it be remembered) and four of her chief ministers. Elizabeth had been excommunicated by Pope Pius V, and his successor, in 1580, laid down "that there is no doubt that whosoever sends her out of the world with the pious intention of doing God service, not only does not sin but gains merit." The French Catholics, led by Mary's cousin, the Duke of Guise, said openly that they were prepared to pay handsomely for Elizabeth's murder. No wonder poor old Bishop Mandell Creighton, writing his *Life of Queen Elizabeth* in the 1890s, once piously suggested sixteenth-century history should be banned since no one could study it without imperiling his soul!

Against this background, it was not surprising that the Catholic party in England were prepared to overlook the Darnley murder and still regard Mary as a possible queen. During her long imprisonment various conspiracies were concocted on her behalf. One involved the Duke of Norfolk, the senior English Catholic layman. He apparently planned to marry her, and become king consort himself. He never set eyes on her and was perhaps unaware he was a good six inches shorter. The plot was discovered, and he paid for his foolishness with his head: yet another of Mary's lovers who came to a sticky end. Elizabeth did not want the odium of executing a queen, and a relative. But as the international situation darkened in the 1580s, and England faced the prospect of an actual invasion by Spain, whose only outcome, if successful, would be the deposition of Elizabeth and her execution—followed by the placing of Mary on the throne—the pressure on the queen by all her Protestant ministers to have done with Mary of Scotland forever, by removing her head, intensified. In 1585 Parliament passed a Safety Bill (27 Elizabeth 1, c.1), with Mary Queen of Scots and her coconspirators very much in mind, and at the same time Walsingham, head of the English security services, contrived to control Mary's correspondence so that he was kept fully informed of her involvement and had copies made of her incriminating letters. Mary duly fell into the trap, and in the so-called Babington Conspiracy she made herself capitally liable under the Safety Bill.

Where Mary was impulsive, Elizabeth had a horror of irrevocable decisions. At first she was unwilling to have Mary tried at all. In the Norfolk plot she had sacrificed Protestant public opinion by having the duke tried and executed while leaving Mary alone. Now, however, she agreed to bring Mary to trial at Fotheringay Castle, beginning October 11, 1585. Babington and company had already been dispatched, some of them with great brutality, but the Protestant bloodlust was not satisfied. For the trial, Elizabeth appointed a special commission, including Lords Lumley and Montagne, both prominent Catholics. The evidence against Mary was so conclusive that Walsingham was able to report: "In the opinion of her best friends, that were appointed commissioners, she is held guilty." She was duly pronounced guilty, though the commission added that her conviction in no way prejudiced the claims of her son James to the throne of England.

Parliament met after the verdict, and expected Elizabeth to sign the death warrant forthwith. On Saturday, November 12, twenty peers and forty MPs attended the queen in her withdrawing chamber at Richmond Palace, to beg her to act. The speaker put their case, and she heard him out. Then she asked them all to draw near to her, and made a carefully thought out but unwritten speech. It was one of those episodes that made Elizabeth Tudor just as much a heroine to one set of admirers as Mary Stuart is to another. The queen said:

> I had so little purpose to pursue [Mary] with any colour of malice, that it is not unknown to some of my Lords here—for now I will play the blab—I secretly wrote her a letter upon the discovery of sundry treasons, that if she would confess them, and privately acknowledge them by her letters unto myself, she never need be called for them into so public question.

Personally, she added, she was willing to forgive even this latest crime. Indeed she would be happy to die if her death could ensure that England got a better monarch: "For your sakes it is I desire to live, to keep you from a worse." She said:

> I am not afraid of death. I have had good trial and experience of this world. I know what it is to be a subject, what to be a sovereign, what to have good neighbours and sometime meet evil-willers. I have found treason in trust. I have seen great benefits little regarded . . . These former remembrances have taught me to bear with a better mind those treasons, than is common to my sex—yea, with a better heart perhaps than is in some men.

As for Mary, she continued,

> In this late Act of Parliament you have laid a hard hand upon me—that I must give directions for her death. But we princes are set on stages, in the sight and view of all the world duly observed. It behoveth us to be careful that our proceedings be just and honourable.

So she must take time to consider—but she assured them all they would "have all conveniency our resolution delivered by our message." Burghley said this speech "drew tears from many eyes."

At a further meeting with parliamentarians on November 24, also at Richmond, she went more deeply into her emotional and moral reasons for hesitating. She admitted that the advice she received was unanimously in favor of Mary's death. But if other means might be found out, she would be "more glad than in any other thing under the sun." Was it not unjust, she asked, that

> I, who have in my time pardoned so many rebels, winked at so many treasons, and either not produced them or altogether slipped them over with silence, should now be forced to this proceeding, against such a person? I have besides, during my reign, seen or heard many opprobrious books and pamphlets against me, my realm and my state, accusing me to be a tyrant. I thank them for their alms. I believe therein their meaning was to tell me news: and news it is to me indeed . . . What will they not now say, when it shall be spread that, for the safety of her life, a maiden queen could be content to spill the blood even of her own kinswoman?

Just to save her own life, Elizabeth said, "I would not touch her." Her object throughout had been not so much to preserve her own life from Mary, but to save the lives of both of them, "which I am right sorry is made so hard, yea, so impossible." She needed more time, and meanwhile she asked them to "accept my thankfulness, excuse my doubtfulness, and take in good part my answer-answerless."

This speech, full of much other material of a highly emotional nature, might have been scripted for her by the young Shakespeare, then just getting started: it is the most personal thing she ever uttered in public. She finally brought herself to sign the death warrant on February 1, 1586, but told her secretary William Davison (as she later asserted) not to send it off without her express order. He either misunderstood or defied her, and sent it off at once, and when she discovered it had been carried out, Davison had to take the blame. He was dismissed from his office and sent to the tower. She even had a plan to hang him, and took legal advice accordingly. He was tried in Star Chamber, punished by an enormous fine and imprisoned "at the Queen's pleasure" (all was subsequently remitted, though she never again employed him). The queen publicly exhibited great rage, real or simulated—probably both—about the execution of Mary. Burghley, her oldest minister—in effect her oldest friend—was also disgraced: "Her Majesty entered into marvellous cruel speeches with the Lord Treasurer, calling him traitor, false dissembler and wicked wretch, commanding him to avoid her presence, and all about the death of the Scottish queen." But this blew over, and Burghley was taken back into favor, remaining there until his death.

Mary was unaware of the convulsion in English public life her execution caused. Might it have given her satisfaction? All her life she was trouble, to her friends, relations, adherents, subjects, enemies and would-be assassins, above all to herself. What a multitude of sorrows whirled around her short life: she was only forty-four when she died, and spent more than half her adult existence in durance, most of it under the threat of execution. She met her death with the greatest possible courage, dignity and grace, on the morning of February 8, 1586, in the great hall of Fotheringay, in front of a thousand people, all members of the English political-Protestant establishment, virtually all men. They were eager to see her. Most had never set eyes on her, had heard of her beauty and were intrigued to see the ravages time had made upon it. She was stripped down to her petticoats, of which she wore many, and after her head was severed, at one stroke mercifully, and held aloft, a small pet dog, concealed all along in her clothes, crept out and whined piteously.

Elizabeth insisted that Mary be buried with royal honors in Peterborough Cathedral, where she remained until her son became king of England in 1603. He directed her remains to be removed to the Henry VII chapel in Westminster Abbey, and had a monument with a recumbent effigy erected over it. In due course he was buried nearby.

Queen Elizabeth I was quite a different heroine than Mary Stuart—a heroine not of romance and tragedy but of real achievement and successful government. She was in the tradition of Henry V. And though she never actually went to war—she hated war and did everything in her power to avoid the struggle with Spain—she did don armor at the time of the Spanish Armada, and thus attired inspected Tilbury Fort in the Thames and made a fighting speech to the troops. In her old age "she kept a rusty sword by her side" and occasionally "in her rage, thrust it into the arras." But she hated war, considering it wicked, nearly always avoidable by sensible rulers, and ruinously expensive. She sought to be a queen of peace and on the whole succeeded. At her death William Camden, the keeper of her records and the foremost historian of the age, paid her a remarkable tribute, which was deserved:

> Though beset by divers nations her mortal enemies, she held the most stout and warlike nation of the English four and forty years and upwards, not only in awe and duty, but even in peace also. Insomuch as, in all England, for so many years, never any mortal man heard the trumpet sound the charge for battle.

How was that achieved? It was a triumph of intelligence and character. Elizabeth learned the basics of sovereignty the hard way. She grew up in the shadow of the tower. Her birth, on September 7, 1533, came as a grievous disappointment and a theological shock to her father, Henry VIII. He was a younger son who had originally been intended by his father to take the post of archbishop of Canterbury, and had received a distinctly religious-oriented education, which he never forgot. He was also highly superstitious. His wife, Catherine of Aragon, had originally been betrothed to his elder brother, and Henry came to believe, when over many years she failed to provide him with a living male child, that he was being punished by God for a marriage he now felt was invalid and incestuous. He studied the canon law of marriage with fierce intensity, especially the relevant texts in Deuteronomy and Numbers, and dismissed the refusal of Rome to dissolve his marriage to Catherine as mere dynastic politics on the part of the Habsburgs, who controlled the pope, Clement VII. His conviction of the theological justice of his case was so complete that he used the occasion to throw off all the authority of the papacy completely, and to make himself head of the English church. He punished capitally those who refused to follow him, such as Bishop Fisher and Sir Thomas More.

Hence Henry divorced Catherine by act of Parliament and married Anne Boleyn with a clear conscience. But when her child was a girl, the king became troubled. He interpreted it as a mark of God's disapproval. On January 27, 1536, Anne gave birth to another child, a boy this time, but stillborn. That finally persuaded Henry that his marriage to Anne was sinful and must be ended. On April 24, he set up a high-powered treason commission to examine Anne Boleyn's conduct. On May 2, Anne was rowed to the tower, entering it, as all prisoners did, by Traitor's Gate. On May 17, five of her supposed lovers were executed. She was pronounced guilty but her execution postponed for two days to allow the Calais headsman to arrive. He was the only expert in the kingdom who knew how to behead with a sword, a less painful method. This was the sole gesture of mercy Henry extended to the woman he had once, presumably, loved. Anne died on May 19. Her marriage had been legally annulled on May 15, and her child Elizabeth bastardized. Her head and body were dumped in an old arrow chest and buried in St. Peter ad Vincula, the tower church.

Elizabeth's thoughts about her mother are unknown. She is recorded as mentioning her only twice in the whole of her life. She had to believe in her mother's innocence to ensure her title to the throne. One of her mother's supposed lovers, Mark Smeaton, a lute player, not only pleaded guilty to committing adultery with Anne, but was established in court as having known her for many years. Elizabeth's half-sister, Mary, when in a bad mood, used to say that she could detect a likeness between Elizabeth and Smeaton. In her last months she told Father de Fresneda, confessor to her husband Philip II, that "she [Elizabeth] was neither her sister nor the daughter of . . . King Henry." Whether Mary actually believed this is another matter. No one else did. Anyone who had dealings with Elizabeth perceived she was every inch the child of that masterful man.

Nevertheless, Mary had every reason to dislike Elizabeth. She was herself a passionate papist, while Elizabeth had been given an education of moderate Protestantism. She was thirteen when Henry VIII died, and was put under the care of his widow Catherine Parr. Catherine remarried, choosing badly: the younger brother of the Lord Protector Seymour, Duke of Somerset. He was a scoundrel, and seems to have made an effort to rape the young teenager Elizabeth, though unsuccessfully. After Catherine Parr died, Seymour proposed marriage to the child, but she referred him to the council. This was just as well, for it shortly emerged that Seymour was at the center of a huge financial and political conspiracy, involving coining, smuggling, dealing with pirates and many other crimes. Seymour was sent to the tower on January 17, 1549, followed by his associates and three members of Elizabeth's household. She nearly went there herself and was closely questioned by Sir Thomas Tyrwhitt, the council's special investigator into the affair, who reported to Somerset that she was a very difficult person to interrogate, refusing to confess anything, "and yet I do see it in her face that she is guilty." He added ruefully: "I do assure Your Grace she hath a very good wit and nothing is gotten of her but by great policy."

Elizabeth escaped the tower this time. Seymour was executed. Bishop Latimer, who attended the beheading, reported: "He died very dangerously, irksomely, horribly . . . whether to be saved or no, I leave it to God, but surely he was a wicked man, and the realm is well rid of him." Elizabeth's teenage comment at the time (as reported later) was: "This day died a man with much wit and very little judgment." Many others followed in his wake. Soon it was his elder brother, Somerset's, turn to fall from power and go to the tower scaffold, and most of his associates, butchered by his rival John Dudley, Duke of Northumberland, who took his place as protector. But then the young king died, and the tragedy of Lady Jane Grey followed, with Northumberland and a score of others going to the block (or in many cases hanged, drawn and quartered). About half the grandees Elizabeth knew as a child were decapitated (or worse). The words that Shakespeare wrote in *Hamlet,* four years from the end of her reign, apply vividly to her childhood and youth:

Of carnal, bloody and unnatural acts,
Of accidental judgments, casual slaughters,
Of deaths put on by cunning and forced cause . . .

That was the world she was brought up in.

Elizabeth bravely and shrewdly kept well clear of the Lady Jane Grey conspiracy. But equally, when Mary came to the throne, Elizabeth refused to become a Catholic and papist, and was told to leave court and live privately. She was inevitably the focus of any Protestant plot, however. Sir Thomas Wyatt, together with a wide-ranging group of high-placed Protestants, planned to make Elizabeth queen, having deposed Mary, on Sunday, March 18, 1554. They were detected, arrested and executed. Under torture, Wyatt refused to implicate Elizabeth, and on the scaffold insisted she was innocent. She was nonetheless held in custody at Whitehall Palace, and sent to the tower on Palm Sunday, having written an agitated and frightened letter to Queen Mary, protesting her innocence. The letter's language is permeated by her horror of the tower. She was well treated because members of the council were aware that Mary might not live long and Elizabeth was next in line of succession. She got the big room on the second floor of the Bell Tower, immediately over More's room. Her room had earlier been occupied by Bishop Fisher, and 150 years later was to hold the Duke of Monmouth, executed for high treason under James II. She was allowed to take exercise along the wall as far as the Beauchamp Tower, provided at least five persons attended her. This gave her a view of Tower Green and the block, then in constant use. She must have seen many executions, probably including that of the firebrand Wyatt—thus hearing him protest her innocence. She probably believed each day might be her last, anxiously inquiring "whether Lady Jane Grey's scaffold has been taken down yet." She got a fright when a new governor of the tower, an ardent Catholic called Sir Henry Bedingfield, was appointed, and a lot of fresh troops in armor joined the garrison. In fact the treatment of the rebellion was on the whole lenient, and there was never much risk of Mary executing her half-sister, a Protestant princess popular with Londoners. In any case the lawyers said there was no evidence against her. Instead, she was taken from the tower on May 19, under escort of a hundred men commanded by Bedingfield, to Woodstock, near Oxford, where she was held nine months. Bedingfield was a punctilious, old-fashioned gentleman who insisted on addressing her on his knees, even while refusing her demands for greater freedom of movement. On April 17, 1555, the house arrest ended, and Elizabeth was summoned to court. In due course, she was permitted to go to her own house, Hatfield. She was twenty-two. For the next three years, she led a curious waiting existence. Queen Mary had failed to produce an heir, and her burning of Protestant heretics was highly unpopular in the towns. She was now the one who was afraid. She doubted her guards and took to wearing armor. She was increasingly ill and everyone waited for her to die. Elizabeth compared the last months of Mary's reign to being "buried alive like my sister." Before she died on November 17, 1558, Mary confirmed Elizabeth's right to succeed her, and the new queen took over power without trouble.

I have recounted the first twenty-five years of Elizabeth's life in a little detail because her experiences were formative in making her the kind of ruler she was. She was the heroine of Protestant England when she ascended to the throne. In the next forty-four years she established herself firmly as a heroine of history, an exemplary ruler and a model for all women called to such a role. Her rule was based upon nine principles, all drawn from her persona life as a much-endangered princess.

First, never marry. She saw how disastrous had proved Mary's marriage to Philip II. A few months before she became queen, she said she wished "to remain in that estate I was, which of all others best liked or pleased me . . . I so well like this estate, as I persuade myself there is not any kind of life comparable to it." She wanted to survive and rule according to her own clear wishes, and she saw that such an aim would be compromised by marriage to whomsoever. For the next twenty years she pretended to entertain ideas of various suitors, but she never had the smallest intention of thus limiting her power. Her virgin state, and her freedom from a marital tie, were the sources of her vast authority.

Second, she picked wise counselors, and stuck to them. Henry VIII had often changed his ministers, so had the boy king and her sister. She chose Sir William Cecil almost as her first act, as her chief minister, and he remained thus until his death, not long before her own. Nearly everyone she used died in her service or outlived her, still in office. She made one or two conspicuous mistakes, notably late in her reign when she raised up the Earl of Essex, believing she could tame him. But on the whole she was a first-class picker, the best of any English sovereign. This is the most valuable of all gifts for a ruler.

Third, she gave the highest priority to finance. At the beginning of her reign she turned to the financier Sir Thomas Gresham to rescue the country from the mess in which Mary's war with France had left it. He was successful, and thereafter Elizabeth and Sir William Cecil (later, as lord treasurer, promoted to Lord Burghley) kept the finances straight. She inherited financial prudence from her cautious grandfather Henry VII, who inspected every line of his exchequer accounts and initialed each page. She was equally meticulous. As a result she never got into a financial mess in forty-four years and her actions were never constrained by lack of cash. Unlike her brother-in-law Philip II, who had infinitely greater means—all the gold and silver of Peru and Mexico—she never overspent. He went bankrupt three times, and had to borrow money at enormous rates of interest, 20 percent, 30 percent or more. Elizabeth could always borrow if she wished, and paid lower interest rates than any other prince of Europe.

Fourth, Elizabeth kept herself straight primarily by avoiding war. In the end she was obliged to fight in France and Ireland, and to resist Spain at sea. But she kept her military liabilities to a minimum. She built no new palaces and spent little on the sixty or so she inherited.

Fifth, Elizabeth believed in opening opportunities for her subjects to create wealth. During her reign, thanks to peace, industry and commerce expanded steadily, agriculture became more efficient, roads, bridges and harbors improved, towns expanded and virtually the entire housing stock of the country was renewed. Next to Holland, England became the richest state in Europe, per capita.

Sixth, Elizabeth rarely innovated. She learned from her father's mistakes that change was always risky, and could be disastrous. Her motto was: if a thing works, leave it alone. Robert Beale, her clerk of the council, in written advice to his successor, noted: "Beware of opinion of being new-fangled and a bringer-in of new customs."

Instead, seventh, Elizabeth sought to improve the usefulness of existing institutions, notably Parliament. She took enormous trouble over Parliament and prepared her speeches to it with great care and cunning, though she made them seem spontaneous. Her relations with MPs were excellent on the whole and were the real key to her financial and legislative success. She reinforced the bonds by elaborate "progresses" through the countryside and towns, which she again prepared carefully, and were highly successful exercises in public relations. There is no doubt she had a genius for public appearances.

Eighth, Elizabeth was moderate in all things—food, drink, entertainment. She took regular exercise. She wisely avoided Tudor medicine and its terrifying drugs, preferring herbal remedies. Except on public occasions, she dressed simply and modestly, often wearing the same simple gown several days running. Her weakness was for sweets, which ruined her teeth. She was moderate in religion, avoiding extremism of any kind and refusing to employ fanatics. Indeed she was the first English sovereign who never gave bishops ministerial jobs.

Finally, Elizabeth believed it was better to take no decision than the wrong one. She practiced masterly inactivity. She learned that doing nothing was often the right thing to do. She was never hasty, precipitous, rash or enthusiastic for action. She thought that the Tudor state was very inefficient and the less it had to do, the better. She had no ideology. She was an empirical conservative. Even today—especially today, perhaps—rulers can learn a lot by studying her life.

Elizabeth's wise governance (or lack of it) nourished a country that became rich in heroes. There were the soldiers and sailors like Drake, or the writers like Shakespeare, or those who combined the two, like Sir Philip Sidney. A good example was Sir Walter Ralegh, with whom I propose to conclude this section on Tudor heroic personalities who lived in the shadow of the axe. He was born in 1554, when Elizabeth had been on the throne six years and was thirty. He came of an old but impoverished Devon family and "spoke broad Devonshire to his dying day." He was the kind of person she liked. "The Queen in her choice," wrote Robert Naunton, an eyewitness, "never took into her favour a mere new man, or a mechanic." But she was always pleased to see a penniless man of ancient stock raise himself by brains and courage, and to help the process. Ralegh crammed a vast amount into his life, starting at fifteen as a soldier fighting with the Huguenots; a student at Oriel, Lyons Inn and the Middle Temple; a poet; a sea captain under his relation Gilbert, where he fought his first naval battle; a commander of a hundred city infantry in suppressing the Munster Revolt. He had twice been in jail for affrays and in his turn had hanged scores of Irish rebels.

The story of his putting his cloak over a puddle so the queen would not get her feet wet is apocryphal but is a poetic truth. Elizabeth would have relished the idea of a poor would-be courtier sacrificing a new garment (probably not yet paid for) for her comfort at a time when his outfit was his only capital. But what seems to have impressed her was not so much his striking looks, though these helped, as his capacity to state his views in the most forthright and eloquent fashion. As Naunton put it,

> True it is, he had gotten the Queen's ear at a trice, and she began to be taken with his elocution, and loved to hear his reasons to her demands. And the truth is, she took him for a kind of oracle, which nettled them all.

Ralegh had all the usual classical book learning but was also fascinated by machinery and science. A century later he would have been a prominent member of the Royal Society, along with Newton, Christopher Wren and other savants. He belonged to a circle that included Dr. John Dee, Thomas Harriot and other empirics, who ran primitive labs and were often accused of necromancy or atheism—and worse. They supplied him with all kinds of information about natural phenomena with which he could fascinate the queen. Typical of his appeal to her was his demonstration, before a skeptical court, of how to weigh tobacco smoke. He recognized in her a fellow intellectual, interested in ideas and with an insatiable curiosity, and also as a woman barred by her sex from activity and adventure in a wider world, who loved to experience vicariously the life of a thinker who was also a man of action.

She employed Ralegh in various capacities. As a colonizer in Ireland, where he was given 12,000 acres in Cork and Waterford (his house is still there). She approved his transatlantic plantation in Roanoke, which he called Virginia after her, but which failed for lack of backup. Ideally she should have made him minister for colonies, to set up what his friend Dee called a "British Empire" in America, which would have advanced the history of the United States by over a generation. But this was precisely the kind of state activism she drew back from—sure to cost more than anticipated and likely to lead to war with Spain. She made him instead warden of the stannaries, where he reorganized the Cornish tin mines, and gave him all kinds of practical jobs in the West Country, especially in naval and defense matters, where he was energetic and practical. He was popular there too.

But he was popular nowhere else, and this made Elizabeth distrust him, since there was usually a good reason for a man's unpopularity. She placed limits on his power, never giving him a major office of state or having him sworn of the council. Why was he hated? Aubrey, repeating the gossip he had heard as a boy, says he "was a tall, handsome and bold man," and "proud, damnably proud." His boldness, in some ways his best quality since it led him to dare the new and the untried, also led him into risky courses, and he often overreached himself. He was the subject of Francis Bacon's famous essay "On Boldness," worth reading if you wish to understand Ralegh. He was also involved in all kinds of skulduggery and the worst rackets of the time—dealing in the confiscated lands of Catholics, for instance, and various monopolies. He took the hatred of himself for granted, even relished it as a sign of success. As he put it in his first published verse:

For whoso reaps renown above the rest
With heaps of hate, will surely be oppressed

The queen did not like this attitude. She thought it hubris. She disliked hypocrisy as much as Ralegh but thought it necessary to oil the wheels of state. Ralegh scorned to employ it. He was tactless and often spoke home truths others took pains to conceal. Thus he told the House of Commons that he "often plucked at a man's sleeve" to prevent him voting in a division, which brought outraged cries of "Oh, oh!" and "Shame!" He was also reckless and rash, unbiddable and disobedient. Defying all the court rules she carefully laid down, he seduced and secretly married one of her maids-in-waiting, Elizabeth Throckmorton, a woman of undisciplined habits and small fortune whom the queen would not have chosen for him. (She was also, to judge by her letters, the worst speller in the country; not that he was much better—his own name was spelled in seventy different ways and he was inconsistent in writing it, though "Ralegh" was the form he used most often.) After this escapade the queen never quite trusted him again. She made him captain of the guard because he ensured an ultrasmart turnout of tall men, personally chosen. It was probably the finest guards' regiment in Europe, and set standards maintained by the British foot guards to this day. He did well materially too, obtaining Sherborne Abbey in Dorset, and a splendid apartment in Durham House, overlooking the Thames. Here was his famous turret-study, "the prospect from which," wrote Aubrey, "is as pleasant perhaps as any in the world." But she never took him into her confidence, any more than she did Francis Bacon, another outstandingly gifted man. The late A. L. Rowse, who knew more about the Elizabethan era than any other historian, used to say to me that this was because Ralegh was "a tremendous liar." It was true. He boasted. He said he could do things that were beyond his power.

This propensity got him in the end. He had been opposed to James I's succession, and when the new king first met him, he said ominously: "I have heard rawly of thee!" Soon, Ralegh was accused of conspiracy, and though he defended himself brilliantly, was condemned to death, his sentence being commuted at the last minute to perpetual imprisonment. He had two rooms in the bloody tower, and was not uncomfortable. But he despaired, and tried to stab himself in the chest with a long knife. It was not until he settled down to writing his *History of the World* (and many other things) that he became composed. He had to be doing things. After twelve years of captivity, he persuaded James to release him to go on an expedition up the Orinoco in what is now Venezuela. He said he could find an enormous gold mine there, and exploit it, without breaking into the king of Spain's dominions. This was a lie, of course. Indeed the whole idea of El Dorado ("The Golden Man") was a fantasy. But James released him and allowed him to do so, on condition he did not fall afoul of the Spanish. The venture was a failure. Ralegh's much-loved son, Wat, was killed, many others perished and a Spanish settlement was burned. When Ralegh returned, brokenhearted, James directed that the original death sentence be carried out. The warrant was drawn up by no less a hand than Francis Bacon's: *De Warranto Speciali pro Decollatione Walteri Raleigh, militis, AD 1618.* He spent his last night not in the tower but in the gatehouse prison at Westminster, and was killed on a scaffold erected in Palace Yard. He was in a jaunty mood and said he would rather die on the scaffold than of a fever. He said he was not afraid to meet God, who had forgiven him his sins. He ate a hearty breakfast, and smoked one last pipe. His death is one of the best documented of all public executions, replete with fascinating detail. Palace Yard was crammed with people, and he had to push his way through them to get to his own scaffold. A man offered him a glass of fine sack wine, and he drank it, cracking a joke: "It is a good drink, if a man might tarry by it." On the scaffold, Ralegh looked around and saluted all he knew, friends and enemies. It was a great turnout of old Elizabethan grandees and new Jacobean celebrities. Among them was John Pym, the great future parliamentarian, and Sir John Eliot. The latter, after witnessing Ralegh's death, changed from a fervent monarchist into a bitter opponent of the Stuarts. The condemned man was allowed to make a tremendous speech. Then he knelt down to pray. He then stood up, gave away his hat and whatever money he had. He shook hands with all the gentry on the scaffold and embraced his friend Lord Arundel, and said: "I have a long journey to go, and therefore I take my leave." He insisted on the executioner showing him the axe, ran his thumb along the blade and said: "This is a sharp medicine but it is a physician for all diseases." His last words were "Strike, man, strike!" His head fell off after two strokes, the lips still moving. The headsman held up the head by its hair but declined to speak the traditional words "Behold the head of a traitor." A great groan went up from the crowd, and a voice cried: "We have never had such a head cut off." The killing of Ralegh was judged a miserable act of cruelty and meanness of spirit even at the time, and drove a hefty nail into the cause of the Stuart absolute monarchy. It gave dignity to a man who had not always possessed it in life, and ensured him a heroic immortality. Did he deserve it? A hero is not judged by ordinary moral standards. He becomes heroic by the image he fixes of himself in our minds. Ralegh's image, brash, bold, proud, brave, adventurous, scintillating and rash, is enormously potent. With all his faults he is an overwhelmingly attractive figure. If he came into the room, now, we would recognize him instantly, and be delighted to see him.

6

IN THE ROAR OF THE CANNON'S MOUTH: WASHINGTON, NELSON AND WELLINGTON

In a book of biographical essays on heroism, there is an inevitable tendency to write too much on warfare. It must be resisted. There is a case, rather, for a categorical objection to military heroes. That was Dryden's view: Homer's heroes were "ungodly mankillers." Swift, in his *Tale of a Tub,* argues that "those Ancient Heroes, famous for their Combating so many Giants, and Dragons, and Robbers were in their own Persons a greater Nuisance to Mankind than any of those Monsters they subdued." Pope denounced generals, conquerors, etc., who were all

> *Much the same, the point's agreed,*
> *From Macedonia's madman to the Swede.*

(By these he meant Alexander and Charles XII.) Johnson too often attacked heroic violence as a contradiction in terms and a source of moral evil. He deplored the use of clichés like "bed of honor," "sinking in death," "joys of conquest" to cover up the horrors, futility and ugliness of war. But Johnson was also a great believer in majority opinion (as a rule) and he recognized that, in all ages, most people respect proven valor in war: "Every man thinks meanly of himself for not having been a soldier." In a delightful letter to Mrs. Thrale's daughter, Queenie, on the subject of military camps, he calls them "perhaps the most important scene of human existence, the real scene of heroick life." He liked the irony of the fact that some of the grandest people in history had passed their lives in tents, "the lowest and most portable accommodation human wit has contrived." He pointed out to the lucky teenagers—imagine getting a letter at school from Dr. Johnson—that warfare was grim, disgusting, uncomfortable and sordid to a unique degree but also unique in its capacity to bring out the virtues of fortitude, endurance and courage.

In any case, as Johnson argued, whether we like it or not, the heroic soldier or sailor will always attract people's admiration, more perhaps than any other kind of great man. So this chapter is devoted to three very different examples of the species. I was tempted to include others. Admiral Blake for instance. I am drawn to him because he came from a part of Somerset where I now have my country house, and which I love. One of twelve sons of a Bridgwater merchant, he is assumed to have gone on sea voyages in his youth, though he first came to public notice, in the civil war, as a colonel in the siege of Bristol, later, as the man who took Taunton and then defended it heroically. Considering how late he came to naval warfare, Blake was astonishingly successful at it, even against such superb professionals as Tromp. He was, moreover, modest, unassuming, chivalrous, truthful and unselfish, a truly good man. But hard to write about, since all the anecdotes prove, on investigation, to be spurious, and all the details false.

With Prince Rupert, one of his great opponents, there is quite another difficulty. He is closely associated with my old Oxford College, Magdalen, having spent much of the war there. A Magdalen man, in fact. He also shared with me a passion for drawing and watercolor painting. He was, indeed, most versatile in his gifts, being as much at home as any admiral and any cavalry commander, clever at science, as a mechanic and inventor, hugely attractive to women, the kind of man who, as a prisoner, gets to bed the jailer's daughter (as he actually did). And yet—there is something about Rupert which makes one unwilling to spend much time in his company. He was saturnine. He exuded gloom. He never seems to have smiled. He was grim even in his cups. He had all the heroic virtues except glamour (and charm). You feel that an evening spent with him listening to his exploits would leave you feeling depressed. So out with Prince Rupert!

Then there is Oliver Cromwell. Carlyle, his biographer, both Homer and Tolstoy (combined) to the hero, rated Cromwell as the ideal man, not only in his *Of Heroes and Hero Worship,* but in his best book, *The Letters and Speeches of Oliver Cromwell.* But that is the trouble. Carlyle made the man his private property. And Carlyle's prose style is fatally catching. You cannot write about Cromwell without falling into those terrible pseudobiblical cadences, rhetorical apostrophes and ironic name tags. To de-Carlyle Cromwell is not impossible—writers have done it. But it requires the kind of vigilance that I find irksome and which for me spoils the huge pleasure of biography. So out with Cromwell, too!

The three men of war I have selected stand toward the end of the eighteenth century and the beginning of the nineteenth, and they participated in the momentous process that ended the ancien régime and launched the modern world. They are all three genuine heroes, unlike the fourth who might have formed with them a quartet: Napoleon. But the last is half hero, half monster, and is best used as a contrast and warning in this context. Washington, Nelson and Wellington make an excellent trio, for the way in which they resembled each other and the many ways in which they differed are alike rich in useful lessons.

George Washington (1732–1799) was a generation older than the other two. By the time Nelson was born in 1758 and Wellington, in 1769, Washington was a senior officer who had been through a world conflict (the Seven Years War) and experienced a shattering defeat as well as victory. He was very much a man of the eighteenth century, an exact contemporary of Haydn and Fragonard. He was unlike Nelson and Wellington in another important respect—he was large, they were small. Size is a key biographical fact, especially in men and above all in fighting men. When I was an office cadet in the British army, the commandant told me: "Beware small generals!" He was a tall man, and he had obviously suffered in his day. The example he obviously had in mind was field marshal Sir Gerald Templar, a small but exceptionally fierce and highly successful general, who won the Malaya war and was the only commander, British, French, American or Russian, to emerge the outright winner in a guerrilla campaign. Small men who emerge at the top of the fighting profession need that extra bit of bellicosity or creative intelligence or sheer courage to compensate for their lack of inches.

Washington was six foot three. That made him enormous in his day. He stood out. He was not particularly burly—not at all bear-like. Elegant, rather, and refined in body, though the Marquis de Lafayette said, "He had the largest pair of hands I have ever seen in a man." Above all, he was tall. This had a powerful influence on his character as a leader of men. He never needed to strut or posture, to push himself, to pull rank, to insist on his power, to demand respect. He got it automatically. His height gave it to him. He was thus what we would call laid back. He was relaxed. He could take subordination and obedience for granted. Statuesque and formidable by nature, he radiated calm and quiet authority.

He was also very strong, and his strength allowed him to do things other generals would never have done. The painter Charles William Peale remembered a moment at Washington's house, Mount Vernon, in 1772, when he and other men were pitching a heavy iron bar, then a popular sport. Washington suddenly appeared and, "without taking off his coat," held out his hand for the missile and hurled it into the air. Peale said "it lost the power of gravitation, and whizzed through the air, striking the ground far, very far beyond our utmost efforts." Washington said: "When you beat my pitch, young gentlemen, I'll try again."

How important that strength and physical superiority was showed itself right at the beginning of his supreme command, when he was with the Continental Army outside Boston. Backcountry regiments from the South for the first time joined the New Englanders in strength, and in Cambridge a regiment of Virginia riflemen, many of them slave owners, dressed "in white linen frocks, ruffled and fringed," met Glover's Marblehead Regiment, many of them free blacks, in "round jackets and fisher's trousers." There was an instant culture war. Insults led to blows, then a fierce hand-to-hand struggle, with "biting and gouging." An eyewitness said that in less than five minutes more than a thousand combatants were on the field. Washington's new army was fighting itself on a scale larger than the battles of Lexington and Concord.

Washington acted immediately, with his black servant William Lee, both on horseback. They rode straight into the middle of the conflict. Washington,

> with the spring of a deer leapt from his saddle, threw the reins of his bridle into the hands of his servant, and rushed into the thickest of the mêlées, with an iron grip seized two tall, brawny, athletic, savage-looking riflemen by the throat, keeping them at arms' length, alternately shaking and talking to them.

The rest stopped fighting and gazed in amazement. Washington did not need to threaten punishment. He quelled, by sheer decisive strength, a riot that might have been fatal to his army.

Washington's height and physique made him a natural leader without any positive effort on his part. By this I do not mean he was in any way passive. He was a vigorous and active man, an early riser, about his business all day, and by no means intellectually idle—he accumulated a library of over eight hundred books, large for the day. His ambition as a boy was to be an affluent country gentleman in the English fashion. He admired all things English (except the treatment of the American colonies, which he thought stupid and ignorant as well as immoral). After his early experiences as a militia officer during the wars against France, he would gladly have taken a regular commission in the British army, in which case we would never have heard of him—he would have served, most probably, in India and the Empire. But he lacked the "interest" to secure one. "Interest," an archetypical eighteenth-century concept, was a word he used often, for he had none. With interest he might have pursued an alternative career, which was proposed to him, in the Royal Navy. Oddly enough this was a possibility also considered by the young Napoleon Bonaparte, born in 1769, in Ajaccio, where he greatly admired the Royal Navy ships he saw visiting Corsican harbors—but he had to reject it for the same reason: lack of "interest."

Instead Washington took the obvious course of going into surveying, thus making himself useful to the grander members of his ramifying family, theoretical owners of hundreds of thousands of acres, largely unfarmed and most of it unsurveyed. This proved of inestimable value. It taught Washington method; keeping of daily, accurate records; map reading and mapmaking; knowledge of the country, especially the interior beyond the Piedmont and the mountains—at the time of the Revolution, he knew more of America than all but a handful of his fellow revolutionaries—and an overwhelming sense of the potentialities of the country, especially in farming.

By various inheritances of his own, and by the much larger inheritance of his wife (18,000 acres plus cash and property), by purchase and reward for service, Washington in time became one of the largest landowners in Virginia, and by his exertions as a farmer-landowner, one of the half dozen richest men in America (his will shows him owning 96 square miles, valued at $530,000 in land and stock alone). He shared the enlightened interest of the best kind of English landowner in scientific farming, bought books on the subject, experimented on his own account and corresponded with experts across the Atlantic. Agriculture was one of the few topics on which he became eloquent:

> I think that the life of a Husbandman is of all others the most delectable. It is honourable. It is amusing. And with Judicious management it is profitable. To see plants rise from the Earth and flourish by the superior skill and bounty of the labourer fills a contemplative mind with ideas which are more easy to be conceived than expressed. The more I am acquainted with agricultural affairs the better I am pleased with them. I can nowhere find so great satisfaction as in those innocent and useful pursuits.

He regarded tidewater farming in Virginia as inefficient and degrading, with no future. It involved slavery. He always owned slaves, and at one time had more than three hundred (mainly belonging to his wife), but he regarded the institution as wrong and incurably wasteful. In the 1760s he farmed over 20,000 acres. Many rich English earls and dukes had no more. Whence the difference in their incomes and his? Because the best English farming was a judicious mixture of arable, pasture and stock raising, all for the market. By contrast Virginia tobacco was bought by London agents, who did the marketing themselves, got the profits and usually had the American planters in their debt. It was a formula for laziness and improvidence. So Washington spent his life switching from planting to scientific farming. He raised wheat, less labor intensive—a skilled plowman could do the work of forty slow-hoeing slaves—but it demanded large numbers of draft animals and they in turn needed large quantities of hay. So he planted corn fodder alongside wheat, raised root crops, forage crops like clover and alfalfa and put out fields to cattle and hogs. They, in turn, and his plow horses, produced manure which he used as fertilizer. He grew peas and potatoes, planted vines and set up fruit and vegetable gardens on all his farms. He detailed seasonal, weekly and daily work procedures, and became expert in tree grafting, sheep shearing, fishing for herring and dragging for sturgeon. He studied with care the markets which America's expanding and multiplying cities provided for every kind of agricultural product.

I stress this side of Washington's life, which is often almost ignored, because it helps to explain why he became such an all-around hero for Americans. The country was about "getting on" by hard work, and for most of the first century of independence, the chief way Americans got on was through the land. Washington was exemplary in this respect, as in so many others. His home, Mount Vernon, which he inherited, embellished, enlarged and made perfect, center and cynosure of a prosperous estate in an idyllic situation, epitomized his taste, industry and hard-earned success.

There is another reason for stressing this side of his life. It made him a revolutionary. Oddly enough, he had a great deal in common with George III, six years his junior. The king also had a passion for agriculture, his chief interest in life—he was known as "Farmer George." Both loved foxhunting. Washington's favorite sport, however, was baseball. It was George's too, though he called it, English fashion, "rounders." (And both were called George after George II.) What Washington chiefly objected to in the proceedings of George III's ministers was not so much "taxation without representation," though he thought it unconscionable, as the efforts of the government to confine the colonies to the seaboard and, in the interests of the Indians, to prevent them from exploiting the interior. To Washington, the interior, and especially the Mississippi Valley, the largest river system in the world with the world's best agricultural land in it, was America's future. He went to war for America's future. He was, in a sense, the first believer in Manifest Destiny.

So much for Washington as a heroic American visionary. As a general he was primarily a strategist rather than a battle commander. He was always outnumbered, up to the final showdown at Yorktown. He had to fight with amateurs, against professionals. He was always short of weapons, uniforms, ammunition, supplies, reliable NCOs and responsible officers. So he could never afford to force a battle or, even, often, to fight a full-scale defensive one. Yet his strategy was clear, intelligent, absolutely consistent and maintained with an iron will from start to finish. He believed he represented the legitimate government of the thirteen colonies, whose traditional powers Britain was trying to usurp. The army he commanded was an entirely legal force, defending its sovereign territory. At all costs the army had to be kept intact and coherent, and prevented from degenerating into a guerrilla force. It did not matter how many skirmishes or even battles he lost, or how often he had to withdraw, or how much territory he had to sacrifice, so long as the army held together. This is why he forced his opponents always to address him as "General Washington" and treat him as a lawful opponent rather than a rebel. His success on this point was not the least of his achievements. So Washington fought a war of attrition. It was his belief that, provided he and his army remained in the field, the financial and human cost of the war to Britain would mount, the political opposition to it would increase and the will to continue it would weaken. He also believed old enemies of Britain—France, Spain, the Dutch—would be tempted to exploit her difficulties. All this proved to be true. Washington's strategy succeeded. And in due course, at Yorktown, the opportunity arose for a decisive stroke. He seized it eagerly and delivered it with speed, resolution and complete success.

He could not have pursued this long-term strategy through many wearisome years without a hard inner core of self-confidence. That in turn had to be based on a firm conviction of the justice of his cause. He was a religious man. There is no record of his ever doing anything he knew to be wrong. He had a conscience, and slavery made it uneasy. His will, liberating his slaves and directing that all be freed after his wife's death, finally relieved him of the burden of guilt on that score. He believed in God, or rather a divine and benevolent Providence. He was not essentially a religious man and took no interest in theology, ritual or clerical matters. He was a Freemason by choice, part of his concern with "interest." But he thought that a general belief in God and the regular practice of a faith was essential to national well-being. These convictions at all stages strengthened his self-confidence in what he was doing as the commanding general of the revolution.

Washington won the war, and that is his primary claim to heroic status. But it is his conduct afterward which is sublime. The example of Oliver Cromwell had shown how difficult it is for a successful revolutionary general to extricate himself from political responsibilities, or to push ambition firmly down. It is to Cromwell's credit that he at least refused the crown. Caesar had succumbed to temptation. Soon, Napoleon was to do likewise, plunging Europe into a decade and a half of ruinous war and causing the death of five million people. It is to George III's credit that he spotted the fresh element of heroism which Washington's victory opened up for him. He asked the American-born president of the Royal Academy, Benjamin West: "What will General Washington do now?" West said he believed he would go back to his farm. George III: "If he does that, he will be the greatest man on earth."

In fact the recall to constitutional service was inevitable. The war had taught him the need for a proper constitution and a strong executive. He took a leading part in steps leading up to the constitutional convention, presided over it, and played a decisive (if unobtrusive, even hidden) part in ensuring it was shaped in such a fashion as to secure rapid and overwhelming approval by the states. He was not only the obvious choice for president but essentially the only one. He had no alternative but to accept the task, unanimously assigned to him by the new nation, of making the Constitution work. And he did it very well. The United States of America has been fortunate in many ways, especially in the magnificent endowment of nature. But not the least of its blessings was the man who first led it to victory, then made the new nation that emerged law-abiding, stable and prosperous, as well as free. This double achievement is without parallel in history.

Yet if Washington is a hero on an objective level, he is not exactly a heroic personality. There is nothing glamorous about him, no glitter, no charisma, no sparkle. He had no small talk (like Wellington). There were long silences at his dinner parties, during which he drummed with his fingers or played with his knives and forks. He was at his best, again like Wellington, in intimate conversation with clever women he could trust to be discreet. Some believed him overrated. His successor, John Adams, thought he was a fortunate booby, who did not really know what was going on. But careful examination of the evidence, which is abundant, both state and personal, usually shows him to be well-informed, industrious and in charge.

Indeed, since Washington kept, from the age of fourteen, every scrap of paper belonging to him—diaries, letters sent and received, accurate and often day-to-day transactions—and saw to it they were carefully arranged and preserved; and since for more than a third of his life he worked in the service of his country, and all that he did officially is recorded in the National Archives on a scale no European country could then equal, his life is the best documented of any spent in the entire eighteenth century, anywhere. All the same, he remains elusive. There is something remote and mysterious about him. No man's mind is so hard to enter and dwell within. It is easy to see that, in achievement, he was a paragon. But a rich or an empty one? A titan of flesh and blood or a clockwork figure programmed to do wisely? I think it is probably best to see him not only as a soldier, or a politician, but as a gentleman farmer, riding about his land from early in the morning to late afternoon, observant, attentive, giving occasional directions, taking notes, doing something he loved and which he knew to be creative.

Thus seen, he was not glamorous. But he was real, solid, comprehensible—and dignified. He was a heavyweight in more ways than one. Benjamin Latrobe put it thus: "He did not speak at any time with remarkable fluency. Perhaps the extreme correctness of his language, which almost seemed studied, prevented that effect." But: "He had something uncommonly majestic in his walk, his address, his figure and his countenance." This is echoed by another eyewitness: "His features are regular and placid with all the muscles of his face under perfect control, though flexible and expressive of deep feeling when moved by emotions. In conversation he looks you full in the face, is deliberate, deferential and engaging. His movements and gestures are graceful, his walk majestic and he is a splendid horseman." "Majestic" is the descriptive word most often used about him. I see him as the Majestic Hero. There we will leave him.

Horatio Nelson (1758–1805) was the epitome of the hero, in its English version. His father was rector of Burnham Thorpe in Norfolk and his mother daughter of a Westminster prebendary. In the eighteenth century, church of England clergymen were exceptionally patriotic, and scores of the best naval officers came from clerical families (e.g., two of Jane Austen's brothers rose to be admirals of the fleet). Moreover, Nelson had "interest." His maternal uncle, Captain Maurice Suckling, took him onboard the *Raisonnable* at the age of twelve, saw him made midshipman, secured a variety of useful jobs for him and, as comptroller of the Royal Navy, had him made sublieutenant the day after he passed the examination, at the early age of nineteen. He saw action on a variety of ships, in a wide range of situations, all over the world, and aged twenty-four, in 1782, was introduced by Lord Hood, his fleet commander, to Prince William, afterward William IV, with the words: "If you wish to ask questions relative to naval tactics, Nelson can give you as much information as any officer in the fleet." Apart from one serious bout of fever, contracted in the attack on San Juan (1780) and four years on half pay (1790–1793), Nelson was in continuous employment all his career, and usually in the thick of things. Given the shortness of his life (he died as vice-admiral at the age of forty-seven), he saw more fighting, on sea, land, upriver, besieging or defending forts, on convoy and escort, on troop landings and evacuations, including four general fleet actions in three of which he was in command, than any other officer in the history of the service.

Given his courage, and insatiable appetite for action, it is amazing he survived as long as he did. He lost an arm and an eye, and sustained wounds or injuries on half a dozen other occasions. He was by nature frail and delicate. He had bouts of seasickness on four occasions we know of. He had a heart condition, Costa's syndrome. He suffered from intermittent bowel trouble. He had bouts of malaria, dysentery, scurvy, flux, septicemia and rheumatism. He lost a lot of teeth and at times kept a hand in front of his face. He fit in well with the contemporary description by the Regency dandy Sir Walter Eliot in *Persuasion,* of the usual appearance of naval officers: "They are all knocked about, and exposed to every climate and every weather, until they are not fit to be seen."

Sir Walter would have disliked too Nelson's "strong Norfolk dialect" and his high-pitched nasal voice. He was always spoken of as "little': "That foolish little fellow Nelson"—Lord Saint Vincent; "That dear little creature"—Lavinia Spencer, wife of the first lord of the admiralty; "My little hero"—Rear Admiral Duckworth. He often looked frail. People thought he pushed himself too hard. Lavinia Spencer again: "The dear little creature puts me into a fidget about his health." She said: "He is exactly like the thoroughbred Mail coach horses that one hears of that go on drawing till they drop down dead, having lost shoes, hoofs and everything." Quite how small he was can never be finally determined. His effigy in Westminster Abbey, supposed to be life-size, is five foot five inches. "Nelson's Spot," a height measured in the old Admiralty Board Room, gives five foot four inches. Calculations from surviving uniforms and other clothes give estimates from five foot four to five foot six inches.

At any rate, he was *seen* as small and delicate. People, and not just women, felt protective toward him. Nelson was vain. He was not vain in Washington's curious, negative way. His greatest ambition was to be thought unambitious—his vanity was to be believed modest to a fault. Nelson was vain in an old-fashioned way. He liked to wear a beautiful full-dress uniform with all his stars. But then, he *knew* he was vain. He was not the son of a conscientious clergyman for nothing. He could analyze himself and recognize his faults, and accept misfortune as God's corrective of them. If he got into difficulties, he never tried to shift the blame to other people. He blamed himself. It was one of his great strengths. A marvelous letter survives in which he describes the worst experience he had in all his naval career, the near loss of his ship *Vanguard*, off Sardinia:

> I ought not to call what has happened to the *Vanguard* by the cold name of accident. I believe firmly it was the Almighty's goodness, to check my consummate vanity. I hope it has made me a better officer, as I feel confident it has made me a better Man. I kiss with all humility the rod. Figure to yourself, a vain man, on Sunday evening at sunset, walking in his cabin with a squadron about him, who looked up to their chief to lead them to glory, and in whom this chief placed the firmest reliance, that the proudest ship, in equal numbers, belonging to France, would have bowed their flags; and with a very rich prize lying by him. Figure to yourself this proud, conceited man, when the sun rose on Monday morning, his ship dismasted, his Fleet dispersed, and himself in such distress that the meanest Frigate out of France would have been an unwelcome guest. But it has pleased Almighty God to bring us into a safe port where although we are refused the rights of humanity, yet the *Vanguard* will in two days get to sea again, as an English Man-of-war.

Nelson's vanity led him to cut a fine figure in battle, wearing all his decorations, a glittering, unmistakable figure, an easy mark for a French sniper in the crosstrees, as proved fatal at Trafalgar. Wellington would never have done that. He rode all over the battlefield, seeing to things himself. But he did not make the enemy's work easy. He dressed in ordinary dark clothes, of a military cut to be sure, but without any distinguishing marks of rank, epaulettes, decorations or anything to attract the eye. He was ubiquitous in action, but inconspicuous. Yet there was method in Nelson's madness.

He wanted his sailors to see him exposed, in the hottest center of danger, on his quarterdeck, throughout the action. The willingness of ordinary seamen to give him their all was one of the secrets of the efficiency of his ships. He had risen at the time of the greatest mutiny in British naval history, at the Nore, and the grievances lingered. Sailors were aware of the huge discrepancy between what they received, when a prize was taken—a few shillings—and what senior officers got, above all, admirals, thousands of pounds. An ordinary gunner once remarked, in Nelson's hearing, that he hoped the likelihood of being hit in action by enemy shot was in the same proportion, as between officers and men, as the distribution of prize money. It was a sobering thought, and Nelson brooded on it. It was his aim to show the men that the danger he ran was at least as great as what they faced—greater, in fact. He was vain of his courage; he *wanted* to be vain of his courage. And his sailors liked his attitude. They too felt protective toward him—and trembled at his exposure to fire. But it was part of his romantic relationship with them, his love affair with the fleet and all who sailed in it. He was an early romantic, a harbinger of the age of romanticism, which began to surge over the West in the first decades of the nineteenth century, Nelson's apogee. To Wellington, of course, all this was nonsense. He was a realist. He insisted his men obey him. He cared little if they had feelings about him. He did not even like them to cheer him: "It comes dangerously close to an expression of opinion." But a navy was not like an army. It was an intimate, highly emotional business, full of sexual analogies, undertones: a ship was "she," "her." Nelson was of a character to heighten these emotions and intimacies.

As a captain-commander, he liked his ship to be perfect, operate like clockwork, crew dedicated, officers picked and as superb in their professional expertise as he was. A young officer testified: "The Captain is a worthy, good man, and much lik'd on board—is much of a gentleman—I don't think there is a ship in the navy better manned throughout." He took more trouble than any other captain to look after the health of his crew, and when a fleet commander, he went into great detail about ensuring correct diet, exercise, variety and climate change. He was a long way ahead of his time in the use of citrus fruits, onions, vegetables, salads and bathing—all forms of cleanliness indeed. He wrote: "The great thing in all Military Service is health, and it is easier for an officer to keep men healthy, than for a Physician to cure them." He was paternal to his budding officers: "I make it a rule to introduce my midshipmen to all the good company I can."

Nelson's relations with his captains came as near to perfection as it is possible in an imperfect world. He trained them in his methods—"the Nelson touch"—precisely so he could delegate the maximum responsibility to them. They were his "band of brothers," "my darling children." His relations with them were romantic—he said his reception by them before Trafalgar was "the sweetest sensation of my life." His methods of delegation, so that captains instinctively knew his mind in action, and signaling was unnecessary, was the most intelligent thing about him, the product of a great mind which was intuitive. It is important to note that Nelson had the capacity to generate joy. His sentry, John Scott, wrote in 1803:

> I have met with no character equal in any degree to his Lordship, his penetration is quick, judgment clear, wisdom great and his decisions correct and decided—nor does he in company appear to bear any weight on his mind, so cheerful and pleasant, that it is a happiness to be about his hand.

With all this, he was by far the most aggressive leader in the entire Napoleonic Wars—more aggressive, if possible, than Bonaparte himself. All his instincts were for action, at the earliest opportunity, on the largest scale, until the enemy was "annihilated"—a favorite word of his. He was not in the least bloodthirsty. He was shipthirsty. He wanted to destroy, incapacitate, but above all capture enemy ships. The enormity of his appetite knew no bounds. He wanted to leave Britain's opponents without a single serviceable warship, leaving her command of the sea absolute. If possible he aimed to blow up enemy ships in port or at anchor, as he had done at the Nile and in Copenhagen, but the next best thing was to grapple, where the huge power of a British broadside and the ferocity of its boarding crews would "do the trick." Hence one of his axioms: "Outmanoeuvre a Russian (contemptuously). But close with a Frenchman." The French could not endure close combat, but on equal terms, and must yield.

He inherited his anti-Gallicanism. He said: "My mother *hated* the French." From the late 1790s until his death, this hatred personified itself in the frustration of Bonaparte's plans. He said: "It matters not at all the way I lay this poker on the floor. But if Bonaparte should say it *must* be placed in this direction, we must instantly insist upon its being laid in some other one." He was one of the first to recognize that Bonaparte could master continental Europe but could not win the world war if the navy did its duty. If he landed 100,000 professional French troops, or even 50,000, safely on the English shore, he might make short work of England's amateur forces of volunteers and militia. That meant Britain could not relinquish command of the narrow seas even for a few hours. That in turn meant the only way security could be ensured, on a permanent basis, was by destroying all France's men-o'-war and those of all her allies. This strategic aim determined his conduct throughout the years leading up to Trafalgar and gave it the obsessive single-mindedness which was its overwhelming characteristic. He was already a hero, but his pursuit of the Franco-Spanish fleet, in the English Channel and the Mediterranean, across the Atlantic and back, and his final "annihilation" of it in the moment of his own death, has a truly heroic quality not often met with in the history of warfare.

It is notable that this lovable, aggressive little man had a way with words. His letters are often outstanding. I like his description of Naples: "A county of fiddlers and poets, of whores and scoundrels"—worthy of his younger contemporary Lord Byron. He had a knack for coining simple phrases which became imperishable: "Kiss me, Hardy." "England expects this day that every man will do his duty." As he lay dying: "I have not been a *great* sinner." "Don't throw me overboard." "Remember me to your father."

He was obsessive as a commander. So it is not, perhaps, surprising he had an obsessive love affair with Lady Hamilton. There is something pathetic about the burning, insensate love of this frail little man for the overwhelming Emma, a good six inches taller and big in proportion. The word often used about her was "colossal." "Exceeding embonpoint." "A tall, imposing figure with the head of a Pallas." "The fattest woman I ever laid eyes on." Their affair excited ridicule among the better-born ladies, most of whom had a soft spot for the admiral. Melesina St. George referred to them as "Antony and Moll Cleopatra." Lady Hardy, having dismissed the widow Nelson married, because he was lonely, as "his commonplace wife," said of the affair: "Lord Nelson has never been in clever, artful women's society and was completely humbugged by Lady Hamilton." She and others did not grasp the nature of her appeal to him. She satisfied his vanity, completely and overwhelmingly. She gave him unequivocal and unrestrained adoration. She had had many men at her feet but made it clear all were nothing compared to the little man. Her adoration reinforced his self-confidence and resolve and so helped to make Trafalgar possible. This should be remembered in considering his weakness. His obsession with her made him behave inexcusably badly to his wife. The details are too painful to recount. On the other hand, the fact that Nelson was a man of weakness, as well as overwhelming strength, adds to his humanity and his ability to inspire affection two hundred years after his death.

It is not easy to conjure up exactly the conduct of a hero in battle, using mere words. A man with descriptive skill is never there at the right time, and those who are tend to be inarticulate. But a certain Lieutenant Parsons was alongside Nelson on the quarterdeck of the *Foudroyant* when he sighted the *Généreux,* one of the

Nile survivors. It is worth quoting in full (I take my text from Edgar Vincent's superb biography, *Nelson: Love and Fame*):

> "Ah! An enemy, Mr Stains. I pray God it may be *Le Généreux*. The signal for a general chase, Sir Ed'ard, [the Nelsonian pronunciation of Edward] make the *Foudroyant* fly!"
>
> Thus spoke the heroic Nelson; and every exertion that emulation could inspire was used to crowd the squadron with canvas, the *Northumberland* taking the lead, with the flagship close on her quarter.
>
> "This will not do, Sir Ed'ard; it is certainly *Le Généreux*, and to my flagship she can alone surrender. Sir Ed'ard, we must and shall beat the *Northumberland*."
>
> "I will do my utmost, my lord; get the engines to work on the sails—hang butts of water to the stays—pipe the hammocks down, and each man shot in them—slack the stays, knock up the wedges, and give the masts play—start off the water, Mr James, and pump the ship.
>
> The *Foudroyant* is drawing a-head, and at last takes the lead in the chase. The Admiral is working his fin [the stump of his right arm], do not cross his hawse I advise you."
>
> The advice was good, for at that moment Nelson opened furiously on the quartermaster at the conn. "I'll knock you off your perch, you rascal, if you are so inattentive. Sir Ed'ard, send your best quartermaster to the weather wheel."
>
> "A strange sail a-head of the chase!" called the look-out man.
>
> "Youngster, to the mast-head. What! Going without your glass, and be d-d to you! Let me know what she is immediately."
>
> "A sloop of war or frigate, my lord," shouted the young signal-midshipman.
>
> "Demand her number."
>
> "The *Success,* my lord."
>
> "Captain Peard; signal to cut off the flying enemy—great odds, though—thirty-two small guns to eighty large ones."
>
> "The *Success* has hove-to athwart-hawse of the *Généreux,* and is firing her larboard broadside. The Frenchman has hoisted his tri-colour, with a rear-admirals flag."

> "Bravo—*Success,* at her again!"
>
> "She has wore round, my lord, and firing her starboard broadside. It has winged her, my lord—her flying kites are flying away altogether. The enemy is close on the *Success,* who must receive her tremendous broadside." The *Généreux* opens her fire on her little enemy, and every person stands aghast, afraid of the consequences. The smoke clears away, and there is the *Success,* crippled it is true, but, bull-dog like, bearing up after the enemy.
>
> "The signal for the *Success* to discontinue the action, and come under my stern," said Lord Nelson; "she has done well for her size. Try a shot from the lower deck at her, Sir Ed'ard."
>
> "It goes over her."
>
> "Beat to quarters, and fire coolly at her masts and yards."
>
> *Le Généreux* at this moment opened her fire on us; and, as a shot passed through the mizzen stay sail, Lord Nelson, putting one of the youngsters on the head, asked him jocularly how he relished the music; and observing something like alarm depicted on his countenance, consoled him with the information that Charles XII ran away from the first shot he heard, though afterwards he was called "The Great," and deservedly, from his bravery. "I, therefore," said Lord Nelson, "hope much from you in future."
>
> Here the *Northumberland* opened her fire, and down came the tri-coloured ensign, amidst the thunder of our united cannon.

Thus we leave the little Admiral, at work, gobbling a French man-of-war.

Arthur Wellesley, Duke of Wellington (1769–1852), is a delight to read and write about because, though a copybook hero, he was devoid of pomposity and punctuated his long career with pungent and pithy remarks which cast brief flashes of intense light on himself and his times. It is a thousand pities he never wrote an autobiography, for his dispatches, clear, orderly, always to the point, show he was a born writer. He was a younger son of the Earl of Mornington, of the Anglo-Irish ascendancy, born in Dublin. But he always denied he was Irish: "Sir, because a man is born in a stable it does not make him a horse."

He went to Eton but made no mark. He never said, "The Battle of Waterloo was won in the playing fields of Eton," or anything remotely like it. His mother referred to "My awkward son Arthur, food for powder and nothing more"—an odd remark for a mother, even an Irishwoman. So he went to the French royal military school at Angus, wore a blue uniform with red buttons, and learned equitation. Then he served as an ensign in the 73rd Foot, lieutenant in the 76th and 41st, captain in the 58th and in the 18th Light Dragoons, major and half colonel in the 33rd, all in the years 1787–1793. This was achieved by "interest" and purchase money, as well as diligence. At twenty-seven he was a full colonel. He grew up in the shadow of his grandee elder brother, Irish viceroy and governor general of India, created Marquess Wellesley, whom he described as "A very amiable fellow—when he got his own way." He then had a parallel career in politics, as MP in the old Irish Parliament, and then in three English boroughs, serving as ADC to two Irish viceroys, and as Irish secretary.

His first important active service was in the Low Countries under the Duke of York: "I learnt what one ought not to do—and that is always something." One lesson: "Officers must never be allowed to bring their private carriages on campaign." Another: "The ability to read a map is essential." Wellington became an absolutely reliable map reader: "The secret of success in war is learning what lies on the other side of the hill." He was not so brilliantly imaginative at long-range map reading as Napoleon, who could plan a whole campaign that way, one of the central reasons for his success. But he never made a mistake by faulty map reading, and managed to din the skill into his senior officers, something Napoleon did not always do—one of the reasons he lost Waterloo was because Grouchy was a poor map reader.

Wellington then went to India, where he served for eight years, becoming a "Sepoy General" (a despised term in both the British and French professional armies), winning six major battles, and dozens of minor ones, and successfully conducting five big sieges. He also led a cavalry charge of four regiments—successfully. "It was the only time I led cavalry . . . I learned you cannot trust British cavalry. They will always get you into a scrape." He also learned "the value of long-distance marching. "It is the key to success in India, and anywhere else I believe." (This view would have been endorsed by Napoleon.) He learned to write a good dispatch, immediately after the battle: "My rule always was, to do the business of the day in the day." He learned to be a hands-on commander: "I was always on the spot. I saw everything and did everything for myself." Again: "There is but one way [to run an army] . . . to do as I did—have a hand of iron. The moment there was the slightest neglect in any department, I was down on them. I learnt to control a battle personally and not to trust to subordinates." Of the battle of Argaum: "If I had not been there to restore the battle . . . we should have lost the day." The battle of Assaye was "the bloodiest for the numbers that I ever saw" (he lost 1,584 killed, wounded and missing, 650 of them Europeans). He was in some very tight corners in India, often pitted against enormously superior numbers. He developed an extraordinary coolness in battle, his most notable and valuable characteristic: "The general was in the thick of the action the whole time . . . I never saw a man so cool and collected as he was." Wellington learned from experience, and in those eight years came up against virtually every situation likely to confront a commanding general. He said: "Avoid a night attack if possible." He said he read, and learned from, Caesar's *Gallic Wars,* and that he acted "much as Alexander the Great seems to have done." But most of generalship was "common sense and attention to detail." He said: "When one is strongly intent on an object, common sense will usually direct me to the right means."

Like Nelson, he paid enormous attention to the food, cleanliness and health of the men. His rules for hygiene drill survive, and are admirable. He advised officers: "Drink little or no wine." Indeed, he drank little all his life, usually adding water (the opposite of Alexander). He said: "I cannot tell the difference between good and bad wine." Drinking port after dinner in mess was an abomination to him. His routine was, the moment dinner was over, to call for coffee, drink it quickly, then leave. He pleaded: "I have writing to do." (True.) Quite apart from official writing, returns, dispatches, etc., Wellington was a cleanly man. He bathed every day if possible, and always had a thorough wash. He shaved twice a day if engaged in the evening. From the start, he kept his hair unpowdered and short—no pigtail, queue, etc. He was the first "short back and sides" man in the British army, and gradually his approach became standard, much promoted by his own efforts. Indeed, Wellington, who was involved with the army over sixty years, set his stamp on it in many ways still felt to this day. Officers learned to imitate his taciturnity or brevity. When asked a question he did not wish to answer, or unwilling to make an immediate comment on a piece of information, he had a habit of saying, "Ha!" Gladstone, who heard him thus in 1836, noted approvingly that "Ha!" was "a convenient *suspensive* expression." Asked for advice, as he often was, he could be unforthcoming: "Sir, you are in a devlish awkward predicament, and must get out of it as best you can." (But, when you think of it, this is good advice.) He had a golden rule, though he didn't call it that: "There is only one line to be adopted in opposition to all tricks: that is the steady, straight line of duty, tempered by forebearance, levity, and good nature." He also said: "Always try to keep in good humour with the world."

He returned from India in fine fettle, with a knighthood and £43,000 won in prize money. With a principle too: "I have taken the king's salt, I am *chappawallat*, and have a duty to serve the crown always." He often said, thereafter: "I am the king's indentured servant." On the voyage home he landed briefly on Saint Helena, which he liked, by a curious coincidence staying at the Brians, a house occupied by Napoleon in 1815 while Longwood, his home in exile, was being prepared for him.

Back home, he was spotted by Prime Minister Pitt, before his death, as a winner: "Before appointment, he tells you all the difficulties. Once appointed he says not another word about them." His career in the Peninsular Wars made him the one general in Europe who was persistently successful against the French. He improved his skills as a strategist and, most important, developed highly effective tactics in dealing with France's magnificent artillery. The leading one was getting his infantry to lie down on the reverse slope of a slight hill during the bombardment. This made all the difference to casualties and morale. He also trained his infantry to keep steady, close formation, while keeping at a high rate of fire, during charges by the ferocious French cavalry. These tactics served him well in Spain: one by one, he met and beat nearly all Napoleon's best generals. He was never ashamed to retreat to avoid defeat, and so was never beaten. But some of his Spanish battles were close, and horrifically bloody. He said: "The battle of Talavera was the hardest fight of modern times . . . lasted for two days and a night." He said: "I hate battles . . . you lose your friends and your best men." He begrudged casualties, contrasting himself with Napoleon:

> He could do what he pleased, and no man lost more armies than he did. Now with me the loss of every man told. I could not risk so much. I knew that if I ever lost 500 men without the clearest necessity, I should be brought on my knees to the bar of the House of Commons.

Hence in Spain he fought a war of attrition, designed to exhaust the enemy. He bolstered the Spanish forces and conserved his men. Unlike Bonaparte, he had no grandiose strategy, no talent for blitzkrieg. He was a monument of patience, accumulating small gains. He saw the war for the Peninsula as a long haul, and he was right. It lasted six years and was the most protracted campaign of the whole period. It did Napoleon's military power huge and cumulative, permanent damage, the disaster in Russia coming as a final blow. Because he was a defensive general, Napoleon grossly underrated him. But Wellington, when the odds were in his favor, was quite capable of carrying out a well-prepared attack. In 1813–1814 he fought an offensive campaign, winning repeated victories on a large scale, breaking into France, and so helping to make Napoleon's abdication inevitable.

But this time, Wellington was a duke, a national hero, and along with Castlereagh, Britain's chief negotiator at the Congress of Vienna. In the Peninsula, dealing with the Portuguese and Spanish governments, he had learned a great deal about diplomacy and international politics. He was at one with Castlereagh in seeing peacemaking in severely practical terms, keeping ideological politics out, and judging every settlement by the criterion: Will it last? When Napoleon escaped and raised an army in France, Wellington approved the resolution of the Congress to treat him as an outlaw, and readily signed the text: "By again appearing in France with projects of confusion and disorder [Bonaparte] has placed himself outside the law and rendered himself subject to public vengeance." Being totally unromantic—he had once played the violin, but burned it as soon as he definitely committed himself to soldiering—Wellington had never found Napoleon in the smallest degree appealing. He was a liar and a thief; promises and human lives meant nothing to him. He summed him up:

> His mind was in its details low and ungentlemanlike. I suppose the narrowness of his early prospects and habit stuck to him. What we understand by gentlemanlike feelings he knew nothing at all about. He never seemed himself at his ease, and even in the boldest things he did there was a mixture of apprehension and meanness.

He particularly despised Napoleon's speeches to his troops, something in which he never indulged. They were "false heroics," the "stuff of players."

Waterloo was Wellington's last battle. The powers that be on the Horse Guards made things as difficult as possible. His well-trained infantry from Spain were all back in England or sent to American operations as part of the senseless War of 1812. Many were killed at the battle of New Orleans. Wellington had to make do with a new army put together in haste, together with Dutch, Flemish, Walloon and German troops. (He had nothing against Germans: his long-serving orderly, Bleckermann, was a German hussar; but he had never commanded them.) He was not allowed to choose his own staff. He would never have picked Lord Uxbridge ("A rash fellow") to command the cavalry. Napoleon moved very quickly, as usual, and after damaging Blücher's Prussians, attacked on Wellington's front sooner than expected: "He has humbugged me!" The duke's concern was to calm the Brussels population—panic and the roads would be blocked with refugees, preventing more of his own troops from England joining his army. That is why he attended the Duchess of Richmond's ball, and gave an impression of supreme coolness. (It was held in a big washhouse and was far from sumptuous.)

Though appearing cool, he was working very hard. "I never took so much trouble about any battle." From Thursday, June 15, 1815, to the end of the day of the actual battle, June 18 (and into the nineteenth), he had only nine hours sleep, and was at work for ninety, most of the time on Copenhagen, his wonderful stallion, grandson of Eclipse, the famous Derby winner. This strong, placid mount had already carried him at the battle of Vitoria, the Pyrenees and Toulouse. The duke ate very little during the four days, mostly tea and toast, and a few cold meat sandwiches. On Waterloo day itself, in eighteen hours, he had a cup of sweet tea given him by Kinkaid of the Rifle Brigade, though he ate supper after it was all over and his dispatch written.

On the field of Waterloo, Wellington deployed 67,661 men and 156 guns, against Napoleon's 71,947 men and 246 guns. There were about 30,000 horses in all. All this mass was in an area of three square miles. For purposes of comparison, it was eighteen times the area of the actual fighting at Agincourt. But it was still very crowded and this explains the dreadfully heavy casualties.

Wellington benefited quickly from Napoleon's mistakes. At breakfast, eaten off silver plate, the emperor said the battle would be "*facile comme manger le petit déjeuner.*" He should have attacked as quickly as possible, but there had been a lot of rain and he delayed the attack several times to allow the ground to dry. It was not until 11:25 AM that the French guns opened up. During all this time fresh troops were reaching Wellington and were deployed instantly; more arrived in the course of the battle. Equally important, the hours lost increased the possibility that Blücher's advancing Prussians would join the battle, as they did in the late afternoon. Napoleon's other mistake was to send Grouchy with 33,000 men to head off Blücher. In fact poor map reading meant Grouchy missed Blücher entirely and played no part in the battle.

The duke (unlike Napoleon) was in the thick of the battle throughout, riding backward and forward just behind the front, and giving tactical orders every five minutes or so. At one point, to escape French cavalry, he had to jump Copenhagen into a British square. He gave orders, if possible, verbally, face-to-face with commanders—the safest way. Otherwise they were written on slips of parchment (which could be wiped clean), kept folded in the buttonhole of his waistcoat. (One of them, in pencil, can be seen at Apsley House.)

It is remarkable that Wellington was not hit at Waterloo. "The finger of providence was upon me," as he said. But then he was always lucky. He was hurt only once, at Orthez, in 1814. A French musket ball hit his sword-belt buckle and bruised his thigh badly. At Salamanca, a bullet made a hole in his cloak and another hit his holster at Talavera. He also lost two horses at Assaye. He was often struck by a spent ball—but that was nothing. A fairly spent cannon shot was another matter: it was such a missile which struck Uxbridge, while he was talking to the duke in the closing stages of Waterloo, and cost him his leg. Napoleon was also lucky, being wounded only once in fifty battles. But he lost a lot of his generals: 8 at Eylau, 12 at Borodino, 16 at Leipzig (plus 102 wounded in these three battles).

Wellington lost a lot of his friends at Waterloo. Their deaths sickened him. He said: "There is nothing so bad as a battle won except a battle lost." And: "It was a close thing. I was never so near being beat in my life. It would not have done had I not been there." A month later, talking to Lady Shelley, he summed up his whole unheroic attitude to warfare. She said "his eye [was] glistening and his voice broken as he spoke of the losses," and his words were solemn:

> I hope to God I have fought my last battle. It is a bad thing always to be fighting. While I am in the thick of it I am too much occupied to feel anything. But it is wretched just after. It is quite impossible to think of glory. Both mind and feelings are exhausted. I am wretched even at the moment of victory, and I always say that next to a battle lost, the greatest misery is a battle gained. Not only do you lose those dear friends with whom you have been living, but you are forced to leave the wounded behind you. To be sure one tries to do the best for them, but how little that is! At such moments every feeling in your breast is deadened. I am now just beginning to regain my natural spirits, but I never wish for any more fighting.

Wellington was forty-six at Waterloo and had thirty-seven years to live, during which he was the most famous man in Europe. He was obliged to answer endless questions about his campaign and battles but never introduced such subjects. He thought descriptions of battles futile and sure to be misleading: "A battle is like a ball. Everybody sees something. Nobody sees everything." He rejected plans to build him, at public expense, a palace, on the scale of Marlborough's Blenheim: "An absurd idea." Stratfieldsay, a modest and unpretentious country house, was bought instead. He acquired

Apsley House, which until recently rejoiced in the address Number One, London. This has a fine room where the annual Waterloo dinners of old comrades were held. It also displayed (and still does) his trophies. These included a number of masterpieces from the Spanish royal collection. They were looted by Joseph Bonaparte, Napoleon's elder brother, whom he made king of Spain. When his regime collapsed he tried to take them back to France but his coach—part of a cavalcade the duke described as "a travelling brothel"—fell into English hands after the battle of Vitoria. Wellington asked the legimate king of Spain, Ferdinand VII, where he wanted them sent. The king replied that, in view of the duke's incomparable services to Spain, he insisted he keep them. So they are at Apsley House today, and include Velazquez's finest painting, *The Water Seller.*

Another painting from this batch, Van Eyck's so-called *Arnolfini Wedding Picture,* was privately looted by a British cavalry captain. It hung for many years in a private Mayfair house without attracting the slightest attention. Then in 1840 the captain sold it to the National Gallery, London, for £600. If Wellington had known at the time about the theft, he would have had the captain shot. He hesitated a long time before accepting Ferdinand's pictures because he was terrified of being accused of receiving booty. If there was one thing he hated, it was military looting. During the Peninsular War he shot or hanged fifty-two British and twenty-eight non-British soldiers, in nearly every case for looting, compounded (usually) by murder.

Wellington spent many years in the cabinet. In 1828–1830 he was prime minister. He decided, in the winter of 1828–1829, that Catholic emancipation could no longer be safely resisted, and brought in a bill accordingly, carrying it with some difficulty. This was a characteristic piece of realism—to be repeated when he helped Peel to get rid of the Corn Laws in 1846. When the Earl of Winchelsea, a booby, said that Wellington's emancipation of the Catholics was the prelude to reintroducing popery, the duke called him out, and a bloodless duel was fought in Battersea Park. This was his only experience of dueling, another custom he detested. The duke was not a first-class prime minister because he could not delegate. He destroyed his own government in November 1830 by making an intemperate speech, without consulting his colleagues, flatly rejecting parliamentary reform. For three weeks in 1834, when the Whig government was thrown out, and while awaiting the return of Peel from abroad, he conducted the entire government single-handedly, a performance never before or since equaled.

Wellington had a classically unhappy home life. His wife, Kitty Pakenham, who came from the same ascendancy class, "does not understand me." He quarreled with his eldest son too. He did his best to give good advice to the foolish George IV, and got the reply: "Hold your tongue, Sir!"—the only occasion when anyone was ever deliberately rude to the duke. He also resented George IV calling him "Arthur," and claiming, when drunk, to have led a cavalry charge at Waterloo. When the king added, "Was it not so, Arthur?" the duke answered, "As Your Majesty has so frequently observed." He hated boasting, especially on military matters. When Wilkie painted his famous picture *Chelsea Pensioners Listening to the Waterloo Dispatch,* Wellington was told he was expected to buy it so it could hang in the collection at Apsley House. He reluctantly agreed, and was appalled when told it would cost him 1,200 guineas. He paid Wilkie in gold. The painter said: "Quite unnecessary, Your Grace, a cheque will be quite sufficient." "What! And let those young clerks at Coote's Bank know what a damned fool I am?"

Wellington's chief pleasure was the company of clever women (like Washington). His best friend was Harriet Arbuthnot, wife of the head of Woods and Forests, a decent man known as "Gosh!" Harriet was two years older, and exactly the same height, five foot nine, as a sketch of the two some six days before her death from cholera in 1834 clearly shows. They argued on politics, not about aims but tactics, and often quarreled fiercely, then made up. He called her "La Tyranna"; she sometimes called him "The Slave."

It was a relationship not wholly unlike Dr. Johnson's with Mrs. Piozzi. Wellington also called her "Black Cap." The relationship was wholly asexual; General Ayala, close to the duke, called it "*la liaison la plus pure au monde*." Sometimes they rowed in public: they were heard shouting at each other in the Mall. How the passersby stared! He told Lady Shelley: "I am . . . more a slave than ever, and La Tyranna more tyrannical." Her death devastated him and he never quite replaced her—though he was friends with the heiress Mrs. Burdett-Coutts, who wanted to marry him after his wife died. But he declined her audacious proposal. He lived simply to the last. Ayala, who traveled back and forth to Vienna with him, said: "The two English phrases I got to hate most were 'early start tomorrow' and 'cold meat.'" His favorite place was Walmer Castle, his "tied cottage" a warden of the Cinque Ports. In his simple room, still shown just as he left it, is the iron camp bed in which he always slept.

All three of these heroes were painted many times. The results are mixed. The best likeness of Nelson, according to Emma Hamilton, is a life-size waxwork, which has miraculously survived. His column in Trafalgar Square is the most famous monument in the world. North Carolina, wanting a statue of Washington for its capital in Raleigh, commissioned Canova, but he refused to do the president in anything but a toga. The original of this weird effigy was destroyed, but there is a plaster copy in the Canova Museum in Porragno. Of course Washington's colossal head is the first in the line on Mount Rushmore. Wellington has nothing like that. But he was painted, brilliantly, by a great master, Goya. And his monument by Alfred Stevens, under an arch in the north aisle of St. Paul's Cathedral, is a superlative piece of architectural sculpture. It is little known and rarely visited.

7

TORTURED HEROISM IN A MAN'S WORLD: JANE WELSH CARLYLE AND EMILY DICKINSON

History is crowded with unsung heroines, wives of celebrated but difficult men, who promoted their husbands' interests, put up with their rages, depressions and vanities, comforted them in bad times and remained discreetly in the background during moments of glory. I often think of these consorts of kings and presidents, great creative artists and spiritual leaders when I am writing about their achievements. Lady Longford, wife of the great philanthropist Frank, Earl of Longford, who devoted a long lifetime to piety and good works, once said to me: "What should the wife of a saint be called? A martyr."

The martyred wives of the famous stretch back through time, a ghostly procession of largely unknown ladies, patient in eternity as they were in life. There was, for instance, Eleanor of Castile, queen of Edward I, the great crusader warrior, conqueror of Wales, "Hammer of the Scots," a man of notorious irritability, whose outbursts of bad temper survive in the financial records, thus: "Item, for repairing damage to the crown which the King in his anger did cast into the fireplace." What of the damage to Eleanor? After her death, the king sought to repair it by having a series of superb stone crosses erected at each stopping place of her body on its journey to burial in London. Some of these "Eleanor crosses" survive, and gazing at them today, one is reminded of what the poor woman endured. Or there is Philippa of Hainault, wife of Edward III, another great warrior, who did everything a wife could do to make her famous spouse happy, even putting up with his mistress, Alice Perrers. The effigy of her matronly form survives in Westminster Abbey, and there is a touching passage in Froissart's *Chronicles* describing a repentant Edward shedding tears at her deathbed. Or think of the wife of President Andrew Jackson, plucked from obscurity to marry a supremely difficult, combative and embattled man, only to become (owing to irregularities about their marriage, for which she was blameless) the victim of America's first great age of political pamphlet warfare. The abuse killed her in the end (as Jackson believed), but before that we have a picture in words of the two, no longer young, sitting on the porch together one evening, smoking their long-stemmed pipes.

There are suffering females attracted to great men in the arts too—Sir Joshua Reynolds's sister, for instance, who had much of his talent but who was not allowed to practice her art, as he judged it unseemly. Or J. S. Bach's wife, who bore him a dozen children, remained in the background throughout his long life of perpetual work composing and performing and after his death died in poverty. Mozart's gifted sister met the same fate. Then there is the case of Anne Hathaway, one of the most familiar of all names but, like Pontius Pilate, attached to an unknown person. She left behind the world's most famous cottage and inherited Shakespeare's "second-best bed." Otherwise—women et preteria nihil. Shakespeare, man of genius, actor-playwright-theater owner, ferociously at work until his early death at fifty-two, cannot have been easy to live with. Was Anne a heroine? We do not know.

One case of a writer's spouse we do know about, and that is the vexed case of Jane Welsh Carlyle. I say "know" but that is an exaggeration. Tom Stoppard once wisely observed to me: "No one knows exactly what happens inside a marriage except the two combatants themselves." Jane Baillie Welsh, a doctor's daughter, was born in 1801, married Carlyle in 1826 and died forty years later in 1866, her husband surviving her by sixteen years. Both were prolific letter writers, and striking ones too, and happily many recipients recognized this and kept them. In 1883 appeared the first collection of Jane's *Letters and Memorials of Jane Welsh Carlyle,* annotated by Carlyle himself and edited by James Anthony Froude. In the following century appeared many more collections, and since 1970, a new and complete edition of the letters of both has been in the course of publication by Duke University, North Carolina, and Edinburgh University. At the time I write this, thirty-four volumes have appeared, and since only the year 1857 has been reached, at least twenty more are to come. A project on this scale has revealed many new letters never before published. Well over two thousand from Jane alone now exist, an enormous number for a nineteenth-century woman who was not a queen or an authoress or particularly well known in her own right.

For the first time it has become possible to create, if not a full portrait of the Carlyles' marriage, at any rate an accurate sketch, and thereby to correct a notorious misapprehension. For up till now the Carlyle union has gone down in history as a classic case of a failed marriage between two totally unsuited individuals, each of brilliant gifts, who used them to make the other unhappy. Both have been blamed. As George Meredith put it: "It was so fortunate the Carlyles married each other instead of somebody else. For thereby only two people were made miserable instead of four." The unhappiness of the Carlyles has also been turned into an instrument for condemning the entire Victorian age, and its concept of marriage in particular.

There are two reasons for this picture emerging. The first is the role of Froude. When, after her death, Jane's birthday came around, the sad old widower sorted through her letters to him:

> such a day's reading as I perhaps never had in my life before. What a piercing radiancy on meaning to me in those dear records, hastily thrown off, full of misery, yet of bright eternal love; all as if on wings of lightning, tingling through one's very heart of hearts! . . . I have asked myself, Ought all this to be lost, and kept to myself, in the brief time that now belongs to *me*? . . . As to "talent," epistolary and other, these *letters*, I perceive, equal and surpass whatever of best I know to exist of that kind.

The result was the arrangement of Jane's letters already mentioned. Carlyle also wrote his *Reminiscences,* in which he described his marriage in detail of a kind which was most unusual for the time. Both manuscripts were handed over to Froude, plus many other papers, and he was appointed Carlyle's literary executor. His coexecutives were Forster and Carlyle's brother John, but both died before Carlyle died, leaving Froude in sole charge. Froude was a clever man but also a fool, and an arrogant fool. He published the *Reminiscences* in two volumes in 1881, Jane's *Letters* in 1883 and a full and exceptionally frank biography of Carlyle in four volumes. The first two volumes, *A History of the First Forty Years of Carlyle's Life,* appeared in 1882, and the second two, *A History of Carlyle's Life in London,* in 1884. These eight volumes in all, based upon unlimited access to the sources, appeared to give an authentic picture of the Carlyles' life together based on Froude's quite mistaken view that Carlyle was Oedipus and Jane Iphigenia, and their marriage a classic Greek tragedy. Froude carved this story in stone—or set it in concrete.

Froude's error was compounded by a piece of mischief making by the late Victorian journalist-fantasist Frank Harris, author of a notorious autobiography, much of it fiction, *My Life and Loves*. In 1910–11 he published, in the *English Review,* "Talks with Carlyle," in which he alleges that the elderly writer, during a walk in Hyde Park, admitted he was sexually impotent, that he and Jane had never had sex together, and that this was one chief source of her misery:

> The body part seemed so little to me . . . I had no idea it could mean so much to her . . . Quarter of a century passed before I found out how wrong I was, how mistaken, how criminally blind . . . It was the doctor told me, and then it was too late for anything but repentance.

Harris reinforced this by asserting that Jane's doctor, Dr. Richard Quain, told him over dinner at the Garrick Club that when he examined her in the 1860s he found she was *virgo intacta.*

There is no objective evidence that either of these highly improbable conversations ever took place. It is almost certain Harris invented both. Credence was attached to them only because Froude had prepared the way by his "tragic marriage" story. Of course! Here was the missing clue! It is true Jane never had a child. That was not uncommon among Victorian couples. No doubt it was a source of sadness to her. But now we knew why—Carlyle had refused to have sex with her, being unable to do so. He was impotent. It was just like Ruskin and Effie, only worse, for Jane was too noble to seek a divorce on the grounds of non-consummation. But no wonder she hated Carlyle in consequence and sought revenge!

It is all nonsense. There is no evidence Carlyle was impotent. Or that he refused to have sex with his wife. Or that she was *virgo intacta*. Their sex life was, in all likelihood, normal. We might know more about it if we had the correspondence between Jane and Geraldine Jewsbury, the novelist, who at times was Jane's best and intimate friend (they also quarreled). They certainly discussed intimate matters in their letters. Unfortunately a Mrs. Alexander Ireland, who edited the Jewsbury letters, cut them ruthlessly and then destroyed the originals. And Jane's letters to Jewsbury were all destroyed by the recipient. Jane wrote over a thousand letters to Carlyle which sometimes made numerous criticisms of his behavior, but none involve his sexuality or lack of it. As for her taking revenge on him, in nearly ten thousand letters of his to her, "there is never a word of counter-criticism."

So where does this leave us? What was their marriage really

like? And was Jane a heroine or a martyr? The answer, provided by Carlyle's best modern editors, Kenneth J. Fielding and David R. Sorensen, is that "The Carlyles' lives were less a classical tragedy than an extensive patchwork rug." And Jane was too self-reliant, combative and good tempered (as well as bad tempered) to be a martyr, though she has strong claims to heroine status. She married Carlyle after much hesitation, and repeated refusals, well knowing what she was taking on. She came from a higher social class. She was an only child (one of the keys to the whole story). She was beautiful, vivacious, witty, clever and highly articulate, "the best match in her parish." She married Carlyle because she wanted to marry a genius, and thought he was one. And she was right. She never had any disappointment on that score, and the letters in which she congratulates him on *Frederick the Great,* his longest, most difficult and agonizing (to both of them) book, is a superb paean of praise, the kind of letter an author longs to get from his wife, and rarely does. He in turn, when he finished the manuscript of *The French Revolution,* gave it to her and said:

> You have not had for a hundred years any book that comes more direct and flamingly from the heart of a living man—do what you like with it, you angel.

Their life was hard. Her life with him was doubly hard. For some years they lived on a remote farm belonging to her, where he learned his craft as a writer. She was lonely, and there was no way in which she could help him. Writers have to learn to write the hard way—by themselves. They were poor, and she had to do menial work. He was an exceptionally difficult man. He once described himself as sitting, surrounded by his thoughts "all inarticulate, sour, fermenting, bottomless, like a hideous enormous Bog of Allan—and I have to force and tear and dig some kind of main ditch through it." Offered a job in an office, he replied: "I admire your faith that a hungry polar bear, reduced to a state of dyspeptic digestion, may safely be trusted tending rabbits."

Jane had to live with this hungry, dyspeptic polar bear, sitting in the Bog of Allen of his thoughts, and it was not easy. His upbringing had been abstemious and he was content with the minimal comforts. Hers had been profuse—and she was not. Long after he began to make an ample income, she was obliged to write him a long letter—one of the most brilliant even she ever penned—explaining why she needed more housekeeping money. Being polar bearish, he could not be left to conduct negotiations requiring tact and persuasion. So she interviewed the income tax commissioners on his behalf, and left another scintillating account of it. Bear he might be, but he could be flattered into dancing. Lady Ashburton, wife of the millionaire boss of Baring's Bank, perceived this, and being the leading hostess, drew him into her circle, and flattered him out of his senses. He paid court in response, and Jane was a touch jealous at times. But Froude enormously exaggerated the depth and intensity and duration of her jealousy. In fact her correspondence shows she was pretty intimate with the lady herself, often stayed at her classical palace, the Grange, in Hampshire, with and without Carlyle, and saw the point of the friendship. When the Lady died, and Ashburton remarried, the second wife became one of her closest friends, and she doted on her ladyship. There was no element of black tragedy there.

There are two points about the Carlyles it is vital to understand. Whatever one may think about Carlyle as a writer, social commentator and historian, there can be no doubt about one thing: he was a grumbler of genius. He began well as a young man, looking for literary work, and continued to hone his skills to the end of his long life. Scotland, England, Edinburgh, London, people in general, thousands of individuals in particular, Whig governments, Tory governments, all governments, the upper classes, the middle classes, the working classes—inventions, railways, steamships, the Great Exhibition, foreign policy, Corn Laws, abolishing Corn Laws, progressives, reactionaries—Disraeli, Palmerston, Gladstone—all these and countless other subjects stimulated his grumbling propensities and his curmudgeonly rhetoric. He grumbled about the weather, the cold, heat, rain, snow and wet; about every variety of noise; about his sleeplessness and digestion; about his head, heart, lungs and stomach. He found fault with every book he read, he carped at every visitor, he groused at the newspapers, moaned at politics, grouched at the queen, muttered against the law, bitched at foreigners—and snarled at children. Friends made him choleric, enemies even crosser; he growled at himself too.

Moreover, like his style of writing—the most infectious ever known—his cantankerousness was catching. Jane caught it. She became a grumbler of genius too—and she added a whole new range of subjects, especially servants: idle servants, drunken servants, thievish servants, carpenters, plumbers, bricklayers, men who came to mend stoves, to lay floors, to put up curtains. Postmen. Cabdrivers. Delivery men. Shopkeepers. Guests. Dinner parties. Evening parties. Tea parties. Italian revolutionaries. German historians. Relations. Society women. People who dropped in. She competed with Carlyle's grumbling out of a spirit of self-preservation, then for success, then for the sheer pleasure of it. Again, Carlyle was a hypochondriac, and his ailments, real or imaginary, were a central part of his grumbling mania, especially his gastroenteritis, dyspepsia, headaches, palpitations and all the things that caused them: food, "made dishes," vegetables, meat, fish—anything he ate, in fact; and lack of food, and things he didn't eat. He grumbled about his sleeplessness and the things that caused *that*: street musicians, people who kept chickens, cocks crowing, dogs barking, people shouting or singing or just walking the streets. His hypochondria was also catching, and Jane caught that as well. Gradually as she processed through life, her headaches became more frequent, lasting and painful; she caught cold more often, and her colds took longer to go away. Events prostrated her, servants turned her frantic, workmen drove her to the brink of madness, suicide, total misery. The Carlyles competed more and more persistently, industriously and imaginatively in the oratory of their ailments, the epistolary glory of their discomforts—and doubtless in their daily, unrecorded exchanges still more. Two first-class hypochondriacs, both grumblers of world quality, living in close proximity or, if apart, exchanging ferocious letters daily—huge, quotidian salvos—it is a formula for something rich and strange. But not necessarily unhappiness.

Moreover, there was a redeeming quality. Humor. Beneath Carlyle's iracund and peevish crust, there smoldered inextinguishable embers of savage jocosity. The humor was black. It was not far removed from rage. But it was strong, and tongues of humorous fire licked over the glowing coals of his complaints, sometimes leaping out high or exploding in a shower of infernal sparks. He could make Jane laugh, and did, often. She too had a powerful streak of humor. She was a natural raconteur. She could take a character coming to their Chelsea house and transform him into a comic turn, script his utterances with delicate vernacular details and mime his gestures and faces. Or she could turn a tiny domestic incident into an episode in a comic saga: "The Servant Problem," "Workmen Who Won't Work," "The Horrors of Chelsea," and so on. She liked to work up these contes into superb anecdotes that, at the end of the day, when Carlyle emerged from his desperate struggles with his *Frederick the Great,* she could perform for him and see his grim, craggy features relax into a smile, or even detonate a great snort of granitic laughter. She could turn these tales into the stuff of letters too. Dickens believed she could have been a great writer. I doubt that. What we have of prose from her, not much more than scraps, is not impressive. But letters were another matter. She *talked* to the recipient, and became herself in the spluttering pen-shaped words, so that you can hear her firm, rapid, Scotch voice, tumbling over itself with the grim joy of her tale, and shimmering with ironies, sarcasm and comic rage and contempt. Carlyle adored her letters. It is evident from the vast number he wrote to her how vital they were to his peace of mind, his ability to work, his sheer sanity. If one failed to arrive as expected, he was distraught, and he really reckoned on getting one every day he was separated from her.

Jane's heroism emerges powerfully from her letters, both those directed to her God-master-genius-love object, who needed to be amused; and those to intimate friends to whom she could present God obliquely, as monster-tyrant or unfathomable presence. It is hard to convey, summarily, the sheer delight of Jane Carlyle as a letter writer. The effect is cumulative, as the reader begins to grasp her talent for emphasis, her speech rhythm in prose, her virtuosity in persiflage and the patterns of her diatribes and divine grumbles. The selection made by Fielding and Sorensen is an excellent introduction.

In short, it is impossible to do justice to them without endless quotation: they must be read. But here is a characteristic selection of subjects among which she bounced, dodged and darted—the epistolary stand-up comic getting fun, rage and tragic comedy out of a vast range of topics. Thus: Horrible Samuel Rogers. Pimple on her nose at Lady Ashburton's party. Grotesque bare backs in the Ballroom. Cheering up the persecuted Lady Bulwer-Lytton. Pentonville Prison—ugh! Great Exhibition disaster. Dreadful life at The Grange. Chaos at Cheyne Walk. Carlyle nearly falling off a mountain. Noises in Chelsea! Ruskins and their divorce—horrors! Carlyle and the Housekeeping Money row. The Hateful Mrs. Gaskell. Buttoning up George Eliot. Bulwer-Lytton rows. "Skittles" putting the Grand Duke in his place. Telling servant how to handle Carlyle. Darwin and the monstrous *Origin of Species*. Dog run over. New baby. Cat's attempts to eat her canary. Donkeys! Noises! Servants! Tea parties and their trials. Dreadful inarticulate young male visitor wastes her time.

Many of Jane's precious (to us!) letters were carelessly discarded or deliberately burned. The Victorians were great creators—and destroyers—of letters. But hundreds of new ones have been found since the Duke-Edinburgh project began forty years ago, and more turn up all the time. I salute Jane the heroic entertainer. But there is a somber theme that runs through her correspondence too: not the difficulties with Carlyle, which she could cope with, and did, but her health. Because in the end, she was not a hypochondriac at all. The problems with her health were real, and they gradually grew worse. Her death in 1866, while out in her carriage, was sudden and instant, but cannot have been unexpected to her. She was a brave soul, and ought to be a heroine among those to whom the tragic comedy of domestic life is the stuff of great drama.

Emily Dickinson (1830–1886), the reclusive poetess of Amherst, Massachusetts, is a heroine, like Jane Carlyle, for many women, especially writers, but for quite different reasons. Her life was a successful struggle against fear, in which she drew nourishment from her creative gifts. She produced 1,775 poems, nearly all short, that have survived, but only seven were published in her lifetime, all without her permission. Though writing poems was a protection against fear, publication of them, to her, was a gross intrusion on her privacy, a rape, or an act like going naked into a drawing room.

The fear was instinctive, natural and lifelong, and sprang from two causes: her religion and her parents. She, like her family, was Calvinist and believed in double predestination. She, like her mother, but unlike her father, believed from an early age that she was "saved," and took Communion, on occasion, without difficulty. Religion pervaded her life, but it is not clear whether she believed in God, at any rate in a benevolent, merciful and loving God. She thought God was unjust and cruel, especially to Moses, in not letting him see the Promised Land: she wrote three indignant poems on this subject (Fr. 521, 179, 1271). She was terrified by the story of Elisha and the forty-two mocking children, whom God killed by the agency of two she-bears. She first learned this in verse from Isaac Watts's *Divine and Moral Songs for Children,* and later commented: "I believe the love of God may be taught not to seem like bears." These were not mythical creatures to her: children had been killed by bears in Massachusetts within living memory. The grizzly was a real monster to her. But she said that the verse in the Bible that most frightened her was "from him that hath not, shall be taken even that he hath." The most important word in her life was "power." She used the word often, and associated it with God, who exercised power for good and evil. Death, especially from tuberculosis and scarlet fever, was common in her age group. Her uncle Asa Bullard edited (in the 1830s) a children's paper, the "Sabbath School Visitor," designed to convert children to religious enthusiasm by emphasizing the imminence of sudden death, physical dismemberment and fatal illness. She later wrote:

I can wade grief
Whole pools of it
I'm used to that
(312)

She was also brought up on the sermons of Jonathan Edwards, which she found horrible. But she did not believe his teaching. In fact, she disbelieved a lot. When she asked: "Who made the Bible?" and was told: "Holy men moved by the Holy Ghost," she compared this answer with what grown-ups said about Santa Claus. She knew God possessed power but she had a skeptical view of the power that parents exercised as God's vicars over her:

They shut me up in Prose
As when a little girl
They put me in the Closet
Because they liked me "still"—
Still! Could they themselves have peeped—
And seen my brain—go round—
They might as well have lodged a Bird
For Treason—in the Pound
Himself has but to will
And easy as a Star
Look down upon Captivity
And laugh—No more have I

But if her parents had no godlike power over her, they could keep her in fear, if only by the contagion of their own fears. Her mother was in a constant quiver of anxiety, always terrified of losing her purse, her sewing, her luggage, her way. She used to say: "When in doubt, don't go out." She was a poor creature: "I never had a mother," said Emily later. Her father was far from a nonentity. Like *his* father, he was much involved, financially and in other ways, in the foundation of Amherst College and its governance, and in the schools associated with it. But his belief that he was not (yet) "saved" led to a fear that he would die thus, and so be separated from those he loved for eternity. This fear dominated his life, and he passed it on. But he filled Emily with physical fears too. She recalled in her fifties that he had taken her to the mill to get grist, and warned her about the powerful mill horse: "Do not get out [of the wagon] and go near the horse, or you will be trampled." She added: "The horse looked round at me, as if to say, 'Eye hath not seen, nor Ear heard the things that I would do to you if I weren't tied.'"

Her father was involved in the state legislature, and so away a lot; and later he served a term in Congress, in the House, and so in Washington. When he was away his family, and especially Emily, were told to observe many restrictive rules, which limited their movements. She was not allowed to go to school if it was cold; no sledding; no play in the snow. Her father had a terror of drafts, of being struck by lightning and of sudden sickness. He and his sister-in-law Lavinia, who lived with them, were both hypochondriacs and valetudinarians, and kept files of remedies for dysentery, cholera and "swallowing forensic acid by mistake," though they often quarreled violently about what to do in an emergency. Edward Dickinson's letters are full of advice about and formulas for health. Staying at home, preferably in bed, was a sovereign remedy. He believed himself to be prone to accidents, and was terrified of what would befall his family if they left the house, especially if he was not there. When away, he told Emily: "Never go out, and lock all doors at all times." His instructions to Emily when she had to go on a visit by train, were disturbing: "When you come home be careful to get out of the car at *Palmer*—don't fall, keep hold of something all the time, till you are safely out—lest they should start, and throw you down, and then run over you." He had a particular fear of her going to a prayer meeting in the vestry, which was a basement: "My positive instruction is that you do not go into the vestry, on any occasion, for any purpose, in my absence. Now don't disregard this. I shall find it out if you do. It is a most dangerous place, and I wonder that anyone will venture into it." He also gave her constant instructions about avoiding snakes. She later wrote:

> I was much in the woods as a little Girl. I was told that the snake would bite me, that I might pick a poisonous flower, or Goblins kidnap me, but I went along and met none but Angels, who were far shyer of me than I could be of them.

She referred to such episodes under the heading "When I was a boy," seeing herself thus, and she wrote of the snake:

> *He likes a Boggy Acre*
> *A Floor too Cool for Corn*
> *Yet when a Boy and Barefoot—*
> *I more than once at Noon*
> *Have passed, I thought, a Whip last*
> *Unbraiding in the Sun*
> *When stooping to secure it*
> *It wrinkled and was gone.*
> (1096B)

Her life was spent between two large houses in Amherst with spacious rooms and big gardens, a touch of Charles Addams perhaps. At Amherst Academy, she read *Paradise Lost,* Edward Young's *Night Thoughts* and Cowper's *The Task*; she did the *Aeneid* in Latin—her textbook survives. She did botany, and geology, which she liked very much, relishing the image of the hills pushing up ("The mountains grow unnoticed"); she also did history, philosophy, algebra, geometry and theology. She was allowed to attend lectures at Amherst College, and had a year (1847) at the Mount Holyoke Seminary. She was described as a twelve-year-old: "a very bright but rather delicate and frail-looking girl; an excellent scholar; of exemplary deportment, faithful in all school duties; but somewhat shy and nervous. Her compositions were strikingly original, and in both thought and style seemed beyond her years, and always attracted much attention in the school and, I am afraid, excited not a little envy." But she did not like school any more than anyone else; thus, at "Noon on Saturday":

> *From all the jails the Boys and Girls*
> *Ecstatically leap—*
> *Beloved only Afternoon*
> *That Prison doesn't keep—*
> *They storm the Earth and stun the Air*
> *A Mob of solid Bliss*
> *Alas—that Frowns should wait*
> *For such a Foe as this*
> (1553)

Her school records reveal tiny, neat handwriting, a sixty-six-page book of pressed flowers. She commented on this:

> *I pull a flower from the woods—*
> *A monster with a glass*
> *Computes the stamens in a breath—*
> *And has her in a "class"*
> (117A)

It is a curious fact that she continued to refer to herself as a "girl" or a "little girl" (as well as a "barefoot boy") well into middle age.

Her year at Mount Holyoke did her no good. She was one of 235 students. They rose at six and every minute of the day had to be accounted for. She worked hard, but the quality of the teaching was poor and she was disappointed, especially by the prominence of religion. She made no lasting friendship. And she was homesick. After that, she never left Amherst except for brief trips. She does not seem to have considered a career. She never came close (like Jane Austen) to getting married. There were men in her life she called her "tutors." The first was Benjamin F. Newton, a student in her father's law firm, who put her on to books and encouraged her to write poetry. He died young. Then there was the Reverend Charles Wadsworth, a Philadelphia preacher, whom she called "my dearest earthly friend," and who taught her that what mattered in religion was not dogma but the feelings. She corresponded with T. W. Higginson of the *Atlantic Monthly* about literary matters, and, in addition to these three, she showed some of her poems to Helen Hunt Jackson, Dr. J. G. Holland and Samuel Bowles. She read continuously. She liked Longfellow's *Kavanagh,* about New England village life, as her annotations in her copy show. She was impressed by *Jane Eyre* in 1851. (The identity of "Currer Bell" was not then known for sure.) Returning the borrowed copy, with a gift of flowers, she wrote: "If all these leaves were altars and on every one a prayer that Currer Bell might be saved—and you were God—would you answer it?" She also wrote, when the identity of Charlotte Brontë was revealed:

> *Oh, what an alternative for Heaven,*
> *When "Bronte" entered there!*
> (146)

Reading *Jane Eyre* coincided with her acquisition of her first and only dog, a Newfoundland, Carlo, and these events seem to have given her self-confidence in writing. Her brother noted: "She is rather too wild at present." A valentine (not for herself to send) survives from this time:

> *Mortality is fatal*
> *Gentility is fine*
> *Rascality, Heroic*
> *Insolvency sublime.*
> (2)

It is worth noting that legal terminology, picked up from her father's remarks about his work, makes a curiously persistent appearance in her poetry, not least references to people going bankrupt—as several members of her family did. It is a disturbing image. Another is the railroad.

I like to see it lap the miles
And lick the valleys up—
(383)

But it is probably impossible to construct an accurate chronological account of Emily's poetic development: the specific evidence is not there. Her poems are on particular themes but not limited as a rule to particular events in her life. They seem of more universal application. This one, for instance, might be categorized "On getting the right book at the right time":

He ate and drank the precious Words
His spirit grew robust—
He knew no more that he was poor,
Nor that his frame was dust—
He danced along the dingy Days
And this bequest of Wings
Was but a Book—what Liberty
A loosened Spirit brings
(1593)

But "He" was not a person.

Commentators have stressed the loneliness and tension of Emily's life in the Addams House/Mansion/Fortress. Her father had black moods, when his daughter had to be silent and walk warily. But Emily called home "a bit of Eden," "a holy thing," and wrote: "Nothing of doubt or distrust can enter its blessed portals." At times it was a lively place. Her father might object to her wide reading. But he did not stop her having "reading parties." There were guests in the home, visitors, noisy gatherings. One eyewitness wrote: "These celestial evenings in the Library—the blazing wood fire—Emily—the music—the rampant fun—the inextinguishable laughter, the uproarious spirit of our chosen—our most congenial circle." Emily at the piano "played weird and beautiful melodies, all from her own inspiration. Oh, she was a choice spirit." These gatherings, however, usually took place not at the Dickinson home, but the house next door, and at a certain point, Emily's father would appear with a lantern to escort her home, though it was only a few yards away.

But she preferred to sit at home writing while the others were out. It gave her, she said, "the *old king feeling.*" It was "the brimful feeling," which social intercourse diminished. It is also true that the more seriously she took her poetry, the more she wrote, the more it moved to the center of her life, the less she wanted to see people.

In the late 1850s she decided she was a "working poet" and that her vocation in life was to write. She began copying her poems from the scraps of paper on which they were originally written, and preserving them in hand-sewn booklets. In the years 1858–1865 she made forty of these compilations, and ten unsewn volumes of them, a total of eight hundred poems. Many others must have been rejected as inferior or not worth preserving. All were kept private. She thought feminine self-respect damaged by publicity. She said to a woman who had published her verse: "How can you bear to print a piece of your soul?" She did not object to criticism but did not need it: she could survive on the occasional word of encouragement:

A little bread, a crust—a crumb,
A little trust, a Demijohn—
Can keep the soul alive—
Not portly—mind!
But breathy, warm—conscious.

Such things as recognition, success, did not apply to her:

Success is counted sweetest
By those who ne'er succeed.
To comprehend a nectar

Requires sorest need—
Not one of all the purple Host
Who took the flag today
Can tell the definition so clear of Victory—
As the defeated—dying
On whose forbidden Ear
The distant strains of triumph
Burst agonised and Clear.
(112A)

She did, however, write a poem about fame, but only one:

Fame of myself, to justify,
All other Plaudit be
Superfluous—An incense
Beyond necessity—
Fame of Myself to lack—Although
My name be else supreme—
This were an Honor horrorless
A futile Diadem.
(481)

This was written during the Civil War, an event which made little impression on Emily, though there is some evidence she felt the war empowered her and certainly 1863 was her most productive year. She was impressed by the dead: "So many brave—have died this year—it don't seem lonely—as it did—before Battle began." She added: "I myself sang off Charnel Steps. Every day life feels mightier, and what we have the power to be, more stupendous" (*Letters,* 298; December 1862). Here is the poem rate for the war years: 1861, 88; 1862, 227; 1863, 295; 1864, 98. Sometimes she wrote two poems in a day.

If the war left her (apparently) unmoved, so did women's rights. When Bowles said women should appear more in public life, she laughed. Then she apologized: "I am sorry I smiled at Women . . . My friends are very few. I can count them upon my fingers and, besides, have fingers to spare. I am gay to see you—because you come so scarcely, else I had been graver!" She was terrified by the possible withdrawal of friendship, but sometimes provoked it: letters from her have been described as "snake bites." The possible—or actual—death of friends sent her into paroxysms of fear:

What shall I do—it whimpers so—
This little Hound within the Heart
All day and night with Bank and Start.
(237A)

Much of her poetry is mysterious, to me, at any rate; and I am skeptical of those who interpret it confidently. There is a fiercely self-destructive poem, calling on fate to "amputate my freckled heart" (Emily was red haired, with freckles on face, hands and arms). She writes of "a Tomahawk in my side." There are sea poems on drowning:

Two swimmers wrestled on the Spar
Until the morning sun—
When one turned, smiling, to the land—
Oh God! the other One!
The stray ships—passing—spied a face
Upon the water borne,
With eyes, in death, still begging—raised
And hands—beseeching—thrown.
(227A)

Is this herself? Who knows. There is another poem:

Title divine—is mine!
The Wife—without the Sign!
Acute degree—conferred on me—
Empress of Calvary!
(194A)

She called her dog, Carlo, "my mute confederate," and his death, after sixteen years, was a blow. The mature, middle-aged Emily is an elusive figure. She had trouble with her eyes, and sometimes banned herself from reading. She had rheumatism. Her life in the big house was one of comparative privilege. Her father's net worth in 1868 was $47,800: a lot. He was the last person in Amherst to be called "squire." She was described as "a little plain woman with two smooth bands of reddish hair . . . in a very plain and exquisitely clean pique and blue net worsted shawl." One article of her clothing survives: a white cotton dress. She rejected fashion. She rejected many visitors too. Those who saw her often used the word "childlike." She presented them with a flower which she called "an introduction." Sometimes she would talk to visitors only through a half-open door, she herself remaining invisible. She might communicate through sybilline notes. She wrote: "Forgive me if I am frightened; I never see strangers, and hardly know what I say." In fact she had considerable power of personality, which she evidently cherished and relished. One visitor commented: "I never was with anyone who drained my nerve-power so much. Without touching her, she drew from me. I am glad not to live near her." Others remembered her as fresh, direct, fascinating. Like a sybil, she had rituals, and was often invisible, but spoke. A visitor trying the handle of the back door was likely to hear her quietly turning the key in the lock. In her last fifteen years she exchanged messages with the locals, who never saw her. Her notes are marked by irony, superiority. They are dry. The claws are never quite sheathed. She often wrote poetry in the cool quiet of the stillroom, making butter and cheese. But if her father tried to force her to go to church, she would take refuge in the cellar, sitting in a rocking chair. Her image of personal liberty was a key in a locked door. She said, performing "her pet gesture of bravado," twisting an imaginary key, "It's just a turn—and freedom." Denying people, not publishing her poems became almost as important as writing them. Once, her outspoken friend Helen Hunt asked her to explain an enigmatic poem Emily had given her. There was no answer. Hunt flared out: "You are a great poet, and it is a wrong to the day you live in, that you will not sing aloud. When you are what men call dead, you will be sorry you were so stingy."

She became ill in November 1885 and died the following May. At her funeral Emily Brontë's poem "No Coward Soul Is Mine," one of her favorites, was read. She was said to have "looked thirty, not a grey hair or wrinkle." Most of her letters were destroyed. *Poems by Emily Dickinson* was published in 1890, and quickly went into seven editions. Thereafter followed a long publishing history, and countless books. The best, in my view, is by Alfred Habegger: *My Wars Are Laid Away in Books: The Life of Emily Dickinson,* published in 2001, and I have been guided by it in this account. I have also used the three-volume scholarly edition of her poems published in 1955 and her three-volume *Letters* (1958). Emily's work should not be overrated. Some of it is mere Christmas-cracker stuff, or the kind of smart-wit poetry produced by Dorothy Parker, who had a lot in common with Emily. The best is sublime, moving, unforgettable, magic, and the woman who produced it is undeniably, in her obstinate, tiresome, brave, unflinching, desperate and triumphant way, heroic.

8

TWO KINDS OF NOBILITY: LINCOLN AND LEE

Abraham Lincoln (1809–1865) comes high on the list of enduring popular heroes, at any rate in the English-speaking world. One crude but useful test is the number of times a person, real or imaginary, has been featured in movies. On a list compiled at the end of the twentieth century, Lincoln appeared fifth, with 137 entries devoted to him. The four ahead of him were Sherlock Holmes (211), Napoleon (194), Dracula (161) and Frankenstein (159). Lincoln thus did better than any other real figure except Napoleon. By comparison other real American heroes were far behind: Ulysses S. Grant (50), the successful general Lincoln found at long last, followed by Washington (38) and the "heroic booby," Custer (33).

There is, I think, one word that explains Lincoln's heroic preeminence in the hearts and minds of so many: goodness. He was a good man on a giant scale, a man who raised goodness into a political principle, into a way of public life, and into a code of government activity. And the fact that he came from nothing and nowhere, had little formal education, Christian training or parental guidance—had taught himself morality and made himself a good man entirely by the intelligent cultivation of sound, deep-rooted instincts—makes his character all the more appealing. There is a famous photograph of him taken at the height of the Civil War, when things were going badly for the North. In an attempt to stir up the extraordinarily supine, not to say pusillanimous, General George B. McClellan, Lincoln visited the headquarters of the Army of the Potomac, and was snapped with the entire staff.

These officers were mostly tall for their times, but Lincoln towers over them to a striking degree. It was as though he was a different order of humanity, not a member of a master race but a higher one. And so in a sense he was. There were great men in Lincoln's day—Gladstone, Disraeli, Tolstoy, Dickens, Bismarck, Ruskin and Newman, for example. And in America, major spirits like Sherman, Grant, Whitman abounded. Yet Lincoln seems to me to have been of a different order of moral magnitude, and indeed of intellectual heroism.

Unlike all the others I have mentioned, he appeared to have no real weakness, and in scrutinizing his record it is impossible to point to particular episodes and say "Here he was morally wrong," "There he was inexcusably weak," or "In this case he demeaned himself." Was he, as millions of Americans believed, sent by God—or, as angry men and women in the South were convinced, an emissary of Satan? He seemed at the time, and still seems, somehow superhuman.

All this despite the fact that his life is remarkably well documented and all the evidence has been sifted over and over again. The Lincoln Papers in the Library of Congress alone fill 97 reels of microfilm, and the library also has the Herndon-Weik collection of supporting documents on 15 reels. The big Lincoln bibliography, now over sixty years old, lists 3,958 books on Lincoln, and thousands more have appeared since. Scores of these books are of high quality and great length, and any biographer of the man must, or ought to, read them all.

Yet it is not necessary to delve deep into the documentation and the literature to grasp the essence of the man. Against any episode of the historical background in which he figured, he is both salient and transparent. One reason is physical. Like Washington he towered above the rest. But he also had a striking head and face of a rugged and lined, almost ugly, nobility. That head said it all. Or, rather, it did not say it all, for what needed to be said, Lincoln said or wrote with sublime power. If ever a statesman was a master of words, he was. Perhaps the fact that he was largely self-educated brought him to words with a freshness and sense of discovery so easily lost in the academic pursuit of literary excellence. But there is nothing naive or primitive about Lincoln's use of English. It is simple; but also extremely sophisticated. He chose words not for their grace or glory but for their fundamental accuracy and truthfulness. And in this pursuit of truth he achieved grace and glory as well.

Here is a simple example. Lincoln rose from nothing to the White House through the law. From being a manual laborer in a variety of humble occupations—rail-splitting was the most skilled—he acquired enough book knowledge to set up as a lawyer. He did sufficiently well in his profession to be able to do what it taught him was an overwhelming necessity: to change bad laws by political action. People often say America has too many lawyers. It is even asserted, perhaps justly, that there are more lawyers in America than in the rest of the world put together. That is because it is easier to become a lawyer in America than in any other country, and always has been. If Lincoln had been born in England, France or Germany, it is most unlikely he could have become a lawyer, and we would never have heard of him. As it was, he got ahead. But he was a lawyer with a difference. He became a skilled lawyer but remained a good man. Here is a letter which sums him up:

> Springfield, Illinois
> 21 February 1856
> To Mr. George P. Floyd,
> Quincy, Illinois

Dear Sir,

I have just received yours of 16th, with check on Flagg & Savage for twenty-five dollars. You must think I am a high-priced man. You are too liberal with your money.

Fifteen dollars is enough for the job. I send you a receipt for fifteen dollars, and return to you a ten-dollar bill.

Yours truly,
A. Lincoln

I would like this letter framed and hung on the partners' desks of every law firm in the country. It is brief, simple and embodies action—the enclosed ten-dollar bill—rather than verbal waffle. Not that Lincoln was a softy. In another case, where he had greatly benefited the Illinois Central Railroad Company by a successful piece of law work, they refused to pay what he regarded as a fair fee. He sued them, and got it.

Mr. Floyd, the recipient of the letter, must surely have felt, on reading it, that the writer was a good man. So must Michael Hahn, the recipient of the next letter. The date is March 13, 1864, and Lincoln was writing from the Executive Mansion, as the White House was then called. Louisiana had already been occupied by the North, and Hahn had been installed as governor. Lincoln wanted Hahn to give some blacks the vote quickly, but the letter shows him as a practical statesman, an empiricist, seeking to do good by stealth.

Private
The Hon. Michael Hahn

My dear Sir,

I congratulate you on having fixed your name in history as the first free-state Governor of Louisiana. Now you are about to have a convention which, among other things, will probably define the elective franchise. I barely suggest for your private consideration, whether some of the coloured people may not be let in—as, for instance, the very intelligent and especially those who have fought gallantly in our ranks. They

> would probably help, in some trying time to come, to keep the jewel of liberty within the family of freedom. But this is only a suggestion, not to the public, but to you alone.
>
> Yours truly,
> A. Lincoln

I quote this letter because it struck me, when I saw it in facsimile in an auction catalog (1994), as another example of the way Lincoln's mind worked, and because I can't remember ever having seen it in print. It shows the suggestive, subtle, intuitive side of Lincoln very well, and is characteristically lit up by a striking phrase, "to keep the jewel of liberty within the family of freedom," which—to judge by the appearance of the holograph—occurred to Lincoln as he was writing the letter.

Words, and the ability to weave them into webs which cling to the memory, are extremely important in forwarding political action. This was already true in semiliterate fifth century BC Athens, as Thucydides makes clear, and in republican Rome, as Shakespeare, with his uncanny gift for getting history right, shows brilliantly in *Julius Caesar.* It was even more important in the third quarter of the nineteenth century in America, where most of the population was aggressively literate, and brought up to read and relish key documents—the Declaration of Independence, the Constitution, and the Bill of Rights. That was the way Lincoln himself was brought up (or brought himself up) and he added to the canon two of the first of its documents: the Gettysburg Address and the Second Inaugural Address. Entire books have been written on these speeches, and their evolution. But it seems to me that the key phrases within them came to Lincoln in intuitive flashes, leaping up from a mind that had brooded so long on the nature of political truth and justice, and the frailty of man in promoting them, that it was composed of hot coals from which sparks might be emitted at any instant.

Not that Lincoln was a furnace of rhetoric, or a Man of Destiny or a superhuman force in any way whatsoever. He is not at all the kind of person Carlyle describes in *Heroes and Hero-Worship,* and I have not so far found any reference to him in Carlyle's voluminous correspondence indicating approval. Lincoln was the first to admit that he often, and on the most important occasions, reacted to events rather than directed them. Lincoln was not a will-to-power man but a democrat. He sought to serve the republic, not to impose his ideas upon it. He was a typical American in that he believed passionately in justice, and its embodiment in the rule of law, and in the country's long-term ability always to realize these beliefs in practice. As he put it during one of his debates with Senator Douglas in 1858: "The cause of Civil Liberty must not be surrendered at the end of *one,* or even one *hundred* defeats." But he was an untypical American, and an untypical hero, in that there was a streak of melancholy in his character. He suffered from occasional bouts of depression, and some historians have even argued that he was a lifelong depressive. In 1998, for instance, Andrew Delbanco, in a series of lectures at Harvard, argued that Lincoln's private despair was the engine of his public work: "The lesson of Lincoln's life is that a passion to secure justice can be a remedy for melancholy." In another interpretation, Joshua Wolf Shenk argued in 2005 (*Lincoln's Melancholy: How Depression Challenged a President and Fueled His Greatness*) that it was precisely the depressive nature of Lincoln's mind that gave him a passion for justice in the first place. One needs to take aboard such arguments in considering such a complex man as Lincoln, but it is important not to exaggerate the role that depression played in his public activities. Lincoln was always busy, and almost always busy doing things he wanted to do and which were worth doing. He had time for thought, and no one thought harder than he did, but no time for brooding.

One way in which he was untypical of most Americans was that he did not, strictly speaking, believe in God or, at any rate, a God most of his fellow citizens would have recognized. But he certainly felt there was a guiding providence and that, in the providential scheme, Americans—"the almost-chosen people," as he called them—had an important role to play. America was a pilot state for a better world and if America failed its great test over slavery, the outlook was grim. In order to survive and lead the world, it must remain united: hence the union was the one clear and unassailable principle in Lincoln's worldview, the one point on which he never had any hesitation or doubt.

By comparison slavery was a mere phenomenon. Lincoln thought it an evil—who in his heart did not?—but he refused to see it in inflammatory moral terms. He went out of his way to admit that Southerners were "no more responsible for the origins of slavery than we." He was prepared to live with slavery, at any rate for a time. What he was not prepared to do was see it extended, and that really was the issue on which the Civil War was fought.

Lincoln did not regard blacks as equals. Or rather, they might be morally equal but in other respects they were fundamentally different, and unacceptable as fellow citizens without qualification. He said bluntly that it was impossible just to free the slaves and make them "politically and socially our equals." He freely admitted an attitude to blacks which would now be classified as racist: "My own feelings will not admit [of equality]." The same was true, he added, of a majority of whites, North and South. "Whether this feeling accords with justice and sound judgment is not the sole question. A universal feeling, whether well- or ill-founded, can not be safely disregarded." It is such statements, and many others of a similar nature, which make Lincoln's speeches and writings so riveting. They show that his salient characteristic was candor, a willingness to admit and articulate truth, however inconvenient or unheroic or distasteful or inconsistent it might be.

Lincoln was a pragmatist as well as a democrat. He realized that if the union were to be preserved he must carry a majority, and if possible a big majority, of the people with him. That meant he had to take account of their real feelings. They could be guided and led, up to a point. But they could not be hustled, let alone forced. Fortunately there was not an atom of fanaticism in Lincoln.

The steps by which Lincoln reached his famous decision to emancipate the slaves show his pragmatism and sense of timing at their best. No man was ever more a practical statesman, as opposed to an ideological one. Every aspiring politician, American, British or foreign, should study his career and the way he applied his mind to the fearful problems which confronted him. Lincoln was a strong man and, like most men quietly confident of their strength, without vanity or self-consciousness. There was a little incident toward the end of his life which, to me, is full of meaning. After the fall of Richmond, the Confederate capital, and on the same day Robert E. Lee finally surrendered, Lincoln went to see his secretary of state, with whom he often disagreed, and whom he did not particularly like. Seward had somehow contrived to break both his arm and his jaw. Lincoln found him not only bedridden but unable to move his head. Without a moment's hesitation, the president stretched out at full length on the bed and, resting on his elbow, brought his face near Seward's, and they held an urgent, whispered conversation on the next steps the administration should take. Then Lincoln talked quietly to the agonized man until he drifted off to sleep.

Lincoln could easily have used the excuse of Seward's incapacity to avoid consulting him at all. But that was not his way. He invariably did the right thing, however easily it might have been avoided. Of how many other great men might this be said?

At the same time, in pursuing his overwhelming objective of preserving the union, to him a moral as well as a political necessity, Lincoln showed himself capable of great ruthlessness. He used the power of the presidency to its utmost. Here, Washington, during the Whiskey Rebellion, had set a precedent of a kind, but Lincoln's exploitation of his office, and of the Constitution, was of an altogether different order. He showed that a great republican democracy, once roused to pursue a mighty and righteous object, was capable of a forcefulness, even ferocity, which was both terrifying as well as sublime. This heroic fortitude in enlisting all the power of the union in the cause of right itself transformed the presidency and the nation, and made it possible for Lincoln's strong-minded successors to follow the precedent on the world scene. Lincoln's ruthlessness was the guide for Woodrow Wilson in taking the United States into the First World War and making the peace that followed, for Franklin Delano Roosevelt in fighting the Second World War on the largest possible scale, for Harry S. Truman in using the atomic bomb against Japan, and in mobilizing the free world against Soviet and Communist aggression, for Ronald Reagan in winning and ending the Cold War and destroying the Soviet empire, and for George W. Bush in fighting international terrorism in its homelands.

Lincoln was able to inaugurate this new kind of heroic leadership in American history because he was a new kind of American—someone for whom citizenship of the union was far more important than his provenance from a particular state. In the tremendous events of the Civil War, the central event in American history, Lincoln was not the only hero. The South had to have a hero too. That part could not be played by the president of the Confederacy, Jefferson Davis. He was not a negligible figure. He was in many ways virtuous, consistent, truthful, courageous and always anxious to be just. But he was also narrow and narrow-minded, extraordinarily constricted by his environment and upbringing, no more heroic than a severely blinkered cart horse painfully pulling a heavy wagon on a preordained track to nowhere.

The South, however, found a hero in Robert E. Lee. He was a noble and virtuous man, like Lincoln. But the contrast in their motivations was significant. The two men had quite different ideas about the individual states, which had nothing directly to do with the North-South divide. Lincoln was born in Kentucky, which in the seventeenth century was quite inaccessible beyond the Alleghenies, and was not open to colonists until 1774. It was the post-1800 beneficiary of the Wilderness Road, and the "dark and bloody ground" of Indian warfare. In 1792 it was admitted to the union as the fifteenth state, the first from beyond the mountains, and then only after Virginia ceded title to its theoretical western lands.

After 1840, using the great Ohio River from Louisville, it became a slave market to the South. It had its own Civil War: 30,000 men from Kentucky fought for the Confederacy, against 60,000 for the Union. Lincoln felt no allegiance at all to the state. When he was nineteen, his family moved to Illinois. It had been under French rule until 1763. Then it became part of the Indiana Territory. It was changed to the Illinois Territory in 1809, and was admitted as the twenty-first state in 1818. It had superb agricultural land and was potentially rich in other ways, but it did not attract attention till the Lincoln-Douglas debates of 1858, which first gave it political importance. Lincoln made Springfield his home and Illinois gave him a professional and political career. But it was to the union, "great and strong," to which he felt allegiance and duty, as well as emotional attachment. States' rights were a fact of life of which, as a pragmatist, he took account. But they meant nothing to him spiritually. To Lee it was profoundly different. Virginia really went back to Ralegh's Roanoke colony of 1584, for when a permanent English settlement was established in Jamestown in 1607, Virginia, after Ralegh patroness Queen Elizabeth, was the automatic choice of name. For quite a long time, Virginia *was* the English presence in America, constituting all the land not occupied by Spain and the French. In 1619 the House of Burgesses was founded in Jamestown, the first representative institution set up in the New World, indeed anywhere outside Europe. There are references from this time to "the Colony and Dominion of Virginia"—hence the term "the Old Dominion" applied to it, making it different from all the others, and special. By 1624 it was a royal colony, and by 1641 it was by far the most important, with 7,500 citizens and over 1,000 prosperous farmers and plantations. Under the Commonwealth it was virtually independent and always felt itself to be, and largely was, self-governing. It was, in the 1770s, the natural leader of the rebellion, along with Massachusetts. Virginia's Peyton Randolph was elected president of the First Continental Congress in 1774. Many of the key figures in the creation of the union were Virginian:

not only Washington but also Patrick Henry, Edmund Randolph and John Marshall, the man who effectually created the Supreme Court. Seven out of the first ten presidents were Virginians.

When Lincoln was elected president and the lower South, led by the extremists of South Carolina, seceded, it was by no means clear that Virginia would follow suit. And if Virginia had stuck by the Union, the secession would have become insignificant. Many professional soldiers from Virginia, such as General Winfield Scott and George H. Thomas, made it clear they would remain unionists whatever the Old Dominion decided. It was thought that Lee would take a similar view and Lincoln offered him the command of the new Union army that had to be created. But uncertain what Virginia would do, and determined to follow her for good or ill, Lee declined the appointment. And when in April 1861, by a democratic decision of the whites, Virginia opted for secession, Lee reluctantly went to war on her behalf. As he put it: "I prize the Union very highly and know of no personal sacrifice I would not make to preserve it, save that of honour."

What did he mean by that? Honor was the key word in Lee's life and vocabulary. It meant something very special to him. He came from the old Virginia aristocracy and married into it. His father was Henry Lee III, revolutionary war general, congressman and one-time governor of Virginia. His wife, Ann Carter, was the great-granddaughter of Robert "King" Carter, who owned 300,000 acres and 1,000 slaves. That was the grand side of Lee's background. There was also the dark side.

To put it bluntly, his father became a crook. His claim to be appointed commander-in-chief of the U.S. army was dismissed by George Washington with the euphemistic "lacks economy." He was certainly a big spender, and to finance his tastes he became a dishonest land speculator. Among those he defrauded was Washington himself. He was given the ironic nickname "Light Horse Harry," and eventually went bankrupt, and was jailed twice. When Robert was six, his father fled from his creditors to the Caribbean, and never returned. His mother was left a needy widow with many children. The family's reputation was not improved by a ruffianly stepson known as "Black Horse Harry," who specialized in adultery.

Robert E. Lee seems to have set himself up, quite deliberately, to redeem the family honor by leading an exemplary life of public service. "Honor," a word he pronounced with a special loving emphasis, putting a stress on each syllable, meant everything to him. His dedication to honor made him a peculiarly suitable person to become the equivalent to the South of Lincoln, sanctifying its cause by personal probity and virtuous inspiration. Like Lincoln, though in a less egregious and angular manner, he looked the part. He radiated beauty and grace. He sat his famous warhorse, Traveller, in a statuesquely erect and distinguished posture, the fine stallion too looking the part. Though he was almost six foot, he had small hands and feet, and there was something feminine in his sweetness and benignity. His fellow cadets at West Point called him "The Marble Model." With his fine beard, first tinged with gray, then white, he became in his fifties a Homeric patriarch. Photos of him remind one of the dignified heads of the Roman emperors around the Sheldonian Theatre in Oxford. It is surprising to learn that he was just sixty-three when he died, loaded with tragic honors.

After an industrious youth, he led a blameless life at West Point, and actually saved from his meager pay, at a time when all other Southern cadets prided themselves on acquiring debts. His high grades meant he was accepted by the elite Army Corps of Engineers, in an army whose chief occupation was building forts. His specialty was taming the wild and mighty river which Mark Twain described so unforgettably in *Life on the Mississippi*. Lee served with valor and immense success in the Mexican war of 1846–1848, emerging a full colonel. Then followed posts as superintendent at West Point and cavalry commander against the Plains Indians. In 1859 Lee put down John Brown's rebellion at Harpers Ferry, and reluctantly handed him over to be hanged. Lee owned slaves much of his life but, like most educated Virginians, thought slavery a great evil, which damaged the whites even more than the blacks. (In this he differed profoundly from Jefferson Davis, who actually believed slavery was beneficial to blacks.) Lee joined the South not to preserve slavery but to enable the Old Dominion to preserve its traditional self-government. It was a point of honor, as he saw it.

Lee cannot have been happy with the way the South ran its war. It is important to remember that, whereas Lincoln was able to run a centralized government which at moments amounted to a virtual dictatorship, the South remained a confederacy, with each state retaining elements of sovereignty, not least over its armed forces. It was also handicapped by many other burdens arising from its ideology, not least Jefferson Davis's policy of defending the frontiers of all the Confederate states, making a concentration of its limited armed forces impossible. Up to half were permanently employed on pointless frontier duties. This dispersal of effort went directly against Lee's own view of the strategy the South must pursue if it were to survive. Unlike most people, on both sides, he predicted from the start that the war would be long and bloody. But he grasped that the South had a commitment to the war which many, perhaps most, Northerners lacked. The North had much greater resources of all kinds and must win in the end, unless the South could play upon the North's relative lack of commitment. Its only chance of winning was to engage the bulk of the Union forces in a decisive battle, and win it. This would provoke a political crisis in the North, perhaps force Lincoln's resignation, and open the road to a compromise—which is what Lee had wanted all along.

Lee had an excellent command of tactics as well as a sure sense of strategy, and held high command in some of the bloodiest battles, winning Bull Run, Fredericksburg and Chancellorsville. But he lacked the supreme authority his strategy required. He was not appointed general-in-chief of the Southern forces until February 1865, far too late and only two months before he was obliged to surrender them at Appomattox. Moreover, as a commanding general he had weaknesses. He lacked the killer instinct. Watching his men delightedly chase the beaten and fleeing Unionists at Fredericksburg, he sadly remarked: "It is well that war is so terrible. Otherwise we should grow too fond of it." And he was too diffident to be a great commander. He disliked rows and personal confrontations, inevitable in war if a general is to assert his authority. He preferred to work through consensus. He tended to issue guidance to subordinate commanders rather than detailed, direct orders. At Gettysburg, the gigantic battle he had been waiting for, which gave him a real chance to destroy the main Union army, this weakness proved fatal. Lee's success on the first day was overwhelming, but on the second he did not make it clear to General James Longstreet that he wanted Culp's Hill and Cemetery Ridge taken at all costs. Longstreet provided too little artillery support to Pickett's famous charge. Even so, a few of Pickett's men reached the crest, and it would have been enough, and the battle won, if Longstreet had thrown in all his men as reinforcements. But he did not do so and the battle was lost. Lee sacrificed a third of his men and the Confederate army was never again capable of winning the war. "It has been a sad day for us," said Lee at one o'clock the following morning, "almost too tired to dismount." He added: "I never saw troops behave more magnificently than Pickett's division . . . And if they had been supported as they were supposed to have been—but for some reason, not yet fully explained to me, they were not—they would have held the position and the day would have been ours." Then he paused, and said, "in a loud voice: 'Too bad! *Too bad!* OH, TOO BAD!'"

Lee was a true hero. He insisted on making possible for others the freedom of thought and action he sought for himself. That is a noble aim, but it is not a virtue in a commanding general. "*C'est magnifique, mais ce n'est pas la guerre.*"

After the war, Lee took on the thankless task of running a poor university, Washington College. But he was broken and tired and did not last long. His life was a protracted elegy for the lost South and its noble values, which were perhaps more myth than reality but were nonetheless treasured in his heart. But there was a surprising element of laughter in all his woes. He had a sense of the absurdity of life, as well as its tragedy. When a wartime admirer in Scotland sent him a superb Afghan rug and a tea cozy, Lee delightedly draped the rug around his shoulders, donned the cozy as a helmet and did a little dance while his daughter Mildred played the piano.

He never struck heroic poses. He was modest to a fault, hid from publicity and when in doubt kept his mouth shut. Southern mythmakers have it that his famous last words recalled Gettysburg: "Tell Hill he *must* come up!" and "Strike the tent!" In fact he said nothing. Lincoln too left behind no famous last words. After such lives, what is there to say?

9

CEREBRAL HEROISM: LUDWIG WITTGENSTEIN

Philosophers are not expected to be heroic. And they very seldom are. But two became heroes because they were awarded that status by followers, admirers and public opinion in the field. The first was Socrates, forced to commit suicide by the state for "corrupting youth," then turned into a hero by Plato. The second was Ludwig Wittgenstein (1889–1951). These two were egregious not by accident. They had much in common. Most philosophers try to construct systems of thought of a more or less ambitious kind, into which almost all aspects of human activity can be fitted. Such were, for instance, Plato himself, Aristotle, St. Thomas Aquinas, Kant and Hegel; and, closer to our own time, Martin Heidegger. But Socrates and Wittgenstein did not aspire to system building. Their object was to teach people how to think, and how not to think, and, having thought, how to articulate their thoughts in accurate words. Where the system builders can be seen as positive, constructive and creative, Socrates and Wittgenstein felt obliged—not always but often—to be negative, destructive and countercreative. It was these propensities that made them dangerous, but also, to initiates, heroic.

I only once set eyes on Wittgenstein, in May 1947, when I

was eighteen and an undergraduate of Magdalen College, Oxford. That evening a meeting of the Jowett Society was held at Magdalen, and Wittgenstein dined at high table there, as a guest of our principal philosophy fellow, Gilbert Ryle, then editor of *Mind*. I was dining with two of my mentors, much older than I was, having served through the war, who were completing their degrees, Karl Leyser and John Cooper, both later distinguished historians and fellows of All Souls. As the dons with one or two guests filed in, Cooper exclaimed: "I say—you know who that is with Ryle? "Yes," said Leyser, "it's *Wittgenstein*!" I had never heard the name before, but I was a likely lad, so I said: "Good God!" He was a striking figure too, not very big but handsome and with gleaming eyes, notable even at a distance. What struck me, however, was his open-necked shirt. It was his habitual garb. But I did not know this at the time, and it then seemed extraordinary that anyone should dine at high table without wearing a tie. Wearing one, and gowns, was compulsory even in the body of the hall where I and my friends sat.

But Wittgenstein's life was, from first to last, egregious. He was born in Vienna on April 26, 1889, the same year as two dictators, Salazar of Portugal and Hitler, and two public entertainers, Jean Cocteau and Charlie Chaplin. Hitler was six days older. Wittgenstein's family was Jewish, though baptized. His father, Karl, was an inventive industrialist of astonishing gifts who made himself the richest man in the coal-steel business of the Danube basin. There were eight children, all gifted, all musical, many depressive: two of Ludwig's brothers committed suicide, and Ludwig himself was often tempted to do so. He was the youngest, and spoiled. He was educated at home, in the family palace in Vienna, up to the age of fourteen. A memoir by his sister Hermine describes his precocity. He could make complex machinery from nothing. Aged ten he constructed from scraps of metal and wood a working sewing machine, a "Wittgenstein-Singer," as he called it. The operation was watched with hatred and fear by the family sewing woman.

He became fluent in Latin and Greek and never had any difficulty in speaking and writing languages which interested him, especially English—he was quite capable of lecturing English dons, such as F. R. Leavis, on the meaning of English words. At fourteen he was sent to the Realschule at Linz, where for a year he was a fellow pupil of Hitler, though not (it is said) in the same class. It was there that Hitler learned from the history teacher his basic concepts of pan-German patriotism, and Wittgenstein also, as a young man, was attracted to writers tinged with ultranationalism, anti-Semitism and male supremacy. Both boys, then and after, made powerful, noticeable, penetrating eyes, of the kind usually called "staring." Neither ever mentioned the other.

At the Realschule Wittgenstein received a fine, technically biased education, and he proceeded from it to the Technische Hochschule in Berlin-Charlottenburg, where he studied mechanical engineering. In 1908 he moved to England, where, at Manchester University, he served for three years as a research student in the department of engineering. There he specialized in aeronautics. At all times in his life, Wittgenstein was capable of devising practical machinery or procedures for overcoming physical problems, and when the opportunity arose he did so quickly and with astonishing success. It was the positive and creative side of his nature. At Manchester (and later) he invented, among other things, a jet-reaction aircraft propeller, of the kind developed twenty years afterward by Sir Frank Whittle. He devised a moving part for a helicopter, which later became standard. In the 1920s he designed and built for one of his sisters a house in Vienna, an early domestic essay in the International Modern style, which was notable for its metal fittings, most of which he designed and made himself, as he was dissatisfied with those on the market. They included novel and ingenious door handles, window catches and hinges. During the Second World War, while working as a medical orderly at Guy's Hospital in London, and in a laboratory at Newcastle Royal Infirmary, he designed a much admired machine for recording pulse pressure, and successful methods of mixing pharmaceutical products and dealing with human tissue in dysentery cases.

Wittgenstein might easily have become an outstanding designer of civil and military aircraft, then at the far frontier of engineering. But making highly efficient machinery only appealed to one side, the creative-constructive side of his enigmatic personality. And there was another side. Engineering in Manchester took him deep into mathematics, and his studies in math took on a life of their own. They led him to the work of Gottlieb Frege, who constituted a kind of human bridge between math and philosophy. He passed over this bridge by corresponding with and meeting Frege, who in turn directed him to Bertrand Russell at Trinity College, Cambridge. In 1912 he was admitted to Trinity, where Russell was completing his magnificent *Principia Mathematica* (three volumes, 1910–1913). Russell took him on as a pupil and was soon groaning in mental pain at their marathon sessions of intense argument. Russell wrote: "He was perhaps the most perfect example I have ever known of genius as traditionally conceived, passionate, profound, intense and domineering." In an astonishingly short time, the role of tutor-pupil had been reversed, and Russell found himself learning from Wittgenstein, or rather, trying desperately to resist his pupil's tendency to destroy his work.

Wittgenstein absorbed philosophy effortlessly, though he never read very much, and he then began to pick to pieces the philosophical methods he found current. All his destructive propensities were suddenly aroused, and he found a fierce delight in giving free rein to them. In particular he emphasized the inability of language to bear the weight of meaning philosophers placed on it. Nearly all propositions turned out, on close examination, to be false. Wittgenstein was an engineering philosopher. Successful engineering depends entirely on the degree of accuracy and tolerance needed for the task, and the ability to achieve it. These are not matters of opinion but statistical fact, and attaining them, or not, becomes immediately obvious once the machine or part is put to the test. Wittgenstein, the man of the slide rule, lathe, caliper and machine vise, accustomed to achieving physical perfection in design, and maximum performance, found words, sentences, grammar, syntax—but words above all—slippery, unstable, ambiguous and treacherous. In engineering you can achieve your object to the specific degree of tolerance required, and there is no argument about it. If the thing is impossible, this too becomes unarguably obvious. There is positive and negative certitude. In philosophy, attempts were being made to express with words, to *say* things, which could only be *shown*. It quickly became an axiom of Wittgenstein's, which runs through his entire work like a continuous wormhole, that there are many things we cannot speak about truthfully/accurately, and then we must be silent. But to this there was a coda: "Nothing is so difficult as not deceiving oneself."

Wittgenstein pushed Russell hard, often keeping him up all night arguing. Many years later Russell told me: "With him I came as close as I have ever done to hitting a fellow philosopher." I have forgotten the details of the incident: it concerned an almost-empty lecture room and whether or not there was a hippopotamus in it. Wittgenstein, to Russell's fury, insisted on looking under the chairs to see. Wittgenstein also wrestled with Cambridge's other leading philosopher, G. E. Moore, exponent of what was then called "common sense" philosophy, a type peculiarly antipathetic to Wittgenstein's insistence on the limitations and inadequacy of words. He wore Moore down too, to the point where Mrs. Moore, worried about her husband's health, forbade discussions that lasted more than thirty minutes.

In the years before the First World War, then, Wittgenstein established himself as a ruthless and often destructive philosopher of genius. Other characteristics emerged. He felt spasmodically that philosophy was, or might be, pointless, and that most people attempting it would be better employed doing some form of manual labor. These thoughts could take the form of peremptory and gratuitous advice. "Russell," he said, "give up philosophy." "Moore, give up logic." He sometimes thought he would give it up himself. Then, in October 1913, he went to the other extreme and settled in a remote hut at Skjolden in Sogn, Norway, to escape the distractions and false values of Cambridge, and concentrate fiercely on mathematical logic.

The coming of the European war in August 1914 interrupted this self-imposed exile, and gave some kind of purpose to his life. He joined the Austrian army as a private soldier, refusing all the opportunities open to a man of his wealth and connections to secure an immediate commission. This was characteristic. Moreover, once in the army, he rejected all service outside the front line. He fought mainly on the eastern front against the Russians, exposing himself recklessly to fire and danger, and performing acts of valor which evoked admiration. In particular he repeatedly volunteered for duties as a "forward observer." This involved crawling into no-man's-land with binoculars and a field telephone, and reporting back to brigade headquarters the results of artillery fire. It was, reportedly, the most perilous of all military duties, and the life of those who performed it was short. One may ponder Wittgenstein's motives in thus risking his life, given the suicide record of his family. But there is no evidence he deliberately tried to get himself killed. What the war showed, rather, was his tendency to take up an extreme position on any issue which forced itself on his attention. It is easy to see him becoming a militant pacifist. Russell chose this path, as the war progressed, and in due course went to jail for allegedly inciting British troops to disobey orders. Wittgenstein went to the opposite extreme, and his conspicuous courage eventually led to his being commissioned from the ranks, a rare event in the old imperial army of Austria-Hungary. With the Russian collapse, he was transferred to the Italian front, being taken a prisoner of war when Austria surrendered in November 1918 and her army disintegrated. The Italians held him until August 1919. It must be said that, although Wittgenstein originally chose service in the ranks and the earthy intimacy of the frontline dugout, he did not relish his fellow soldiers' company. He thought them bestial and uncivilized, selfish, disgusting and horrible. He was very clean by nature and habit, and the filth he endured was almost intolerable to him. At Cambridge he leaned to despise intellectuals. In the trenches he found the "common man" equally if not more repellent.

Yet it was during these war years—the period when his schoolfellow Hitler, equally conspicuous for his valor and contempt of danger, formulated his political philosophy—that Wittgenstein added another, and moral, dimension to his thinking, and developed a powerful but occult moral philosophy. It is said to have been the consequence of reading Tolstoy, but I doubt it. There is no evidence that Tolstoy, a loose and imprecise thinker of the kind Wittgenstein despised, left any traces on his written or spoken words. Wittgenstein's reading was never extensive and was always peculiarly eclectic. He was picky, then obsessive. The only Dickens he liked, for instance, was bits of *The Uncommercial Traveller* and a few passages from *A Christmas Carol* and *Nicholas Nickleby*, which he read again and again. He liked reading *Grimms' Fairy Tales,* especially the story "Rumpelstiltskin," whose power was entirely dependent on no one knowing his name. He thought this a transcendental image and it exercised great influence over his thinking. He constantly returned to it. He seems to have had little interest in literature as such. His culture was musical. His family had held regular concerts at their palace, frequented by all the leading musicians of prewar Vienna. One of his brothers became a professional pianist, and when he lost the use of his right hand, Maurice Ravel wrote his Concerto for the Left Hand for him. Wittgenstein was not a performer, but he was a superb whistler, especially of sonata-form music, a skill he may have learned from Gustav Mahler, a friend of his parents, when director of the Vienna opera. Mahler could whistle duets and trios with astonishing virtuosity. Classical whistling seems to have been a much-sought-after accomplishment in pre-1914 Vienna. Hitler was a superb whistler, especially of his favorite, *The Merry Widow,* by Lehár. Wittgenstein liked to whistle an entire symphony, by Schubert or Brahms, with descants and arpeggios, and digressions into the woodwind. But in reading he preferred crude excitement. Between the wars he developed an intense passion for American gangster stories in pulp-fiction magazines such as *The Black Mask.* This was the genre in which Dashiell Hammett and Raymond Chandler learned their literary skills. But Wittgenstein showed no preference for above-average tales of Sam Spade and Philip Marlowe. He liked the repetitive, routine stuff, and the more formulaic the better: what he required was quantity and reliability. During the Second World War it was important to him to receive from friends in America huge periodic bundles of pulp magazines. It was the same with food. No man was ever less interested in gastronomy. But if he found a dish or product to his taste, he would go on eating it every day, with mindless concentration, until a new fancy supervened. All his habits had a tendency toward obsession.

Hence it is not surprising that religion played a poignant part in his life, liable to erupt without warning into obsessive phases, then subside, but remaining always beneath the surface. There was always something monkish about him, practicing poverty, celibacy—but never obedience—and something of the preaching friar too, and the imperious prelate, laying down dogmatic theology ex cathedra. Wittgenstein was never more humble than when sitting barefoot in the bishop's chair. It is not clear whether he ever believed in God, because God, or rather belief in God, raised propositional definitions of intractable complexity. But he once said: "My type of thinking is not wanted in this present age. I have to swim so strongly against the tide . . . I am not a religious man but I cannot help seeing every problem from a religious point of view." When he went to his hut in Norway in 1931, he spent his entire time, or so he said on his return, praying—and did no philosophy. What sort of prayers? And for what intention? We do not know. It is possible to write an entire book about the presence of religion in his life, and its intrusion into his work; indeed, it has been done. But at the end of it all, Wittgenstein's relation to God remains obscure, if indeed it existed at all. He certainly made two points. First, that the idea of God could make a person experience what he called "feeling *absolutely* safe;" the "state of mind" in which you can say "I am safe, nothing can injure me whatever happens." Second, he enjoyed the experience summed up in the phrase, "I wonder at the existence of the world." He relished the idea of "seeing the world as a miracle," the primary miracle of God.

However, Wittgenstein had read St. Augustine's *Confessions,* or at least looked into it. He quoted from it, more than once. He thought that what mattered was not so much belief as conduct. The essence of Christianity, he said (as he had had a Catholic upbringing) was not dogma or even prayer but that "our manner of life is different." He added: "Only if you try to be helpful to other people will you in the end find your way to God." He did not, I think, find it easy to be helpful to people. He valued friendship and could be a passionate friend but also a difficult one. He spoke little about his sexuality—not many people, in those days, made it a topic of conversation. It has been widely assumed he was homosexual. He once arranged to go on holiday with a woman in such a way as to lead her to think he desired to sleep with her. She was prepared to receive him into her bed but he did not attempt to enter it. His friendship with men came closer; in one or two cases obsessive, as one would expect. He used the old-style phrase "lie with" in speaking of such persons. But he also used words like "wicked," "filthy," "disgusting." If he did have sexual intercourse with a man or men on one or two occasions, he repented of it. One academic philosopher told me: "He would seduce young men, whether sexually or intellectually or both I do not know, and then behave brutally to them." This (emotional not physically) brutality reflected guilt.

One way in which he could, however, benefit his fellow humans in a Christian manner was by teaching them to think, clearly and truthfully and in relation to the real world. That meant putting his own thoughts in order. During the war he did this, so that by the end of it, he had completed a tract called *Logisch-Philosophische Abhandlung*, usually known as the *Tractatus Logico-Philosophicus.* He experienced enormous difficulty in writing it, and then difficulty in getting it published. He might have paid for it to be printed but refused, and eventually it appeared in the *Annalen der Naturphilosophie* (1921). An English translation appeared in 1922. Short, pithy, in numbered sections, it is the most memorable philosophical work to appear in the entire twentieth century, anywhere. It is not a book, but closer to a collection of apothegms, linked thematically, though the theme is often subterranean, or even totally absent. No one has ever understood it as a whole, no matter what they may say. Even Wittgenstein, having put it together from bits assembled in his mind over a long period, probably never approved it as a totality. At any rate he gradually came to disbelieve in much, if not all, of it, and his only other major work, *Philosophical Investigations,* published posthumously, is quite different.

There is one important point to be grasped about him. He was quite incapable of writing a *book* about philosophy, just as Karl Marx was incapable of writing a *book* about economics or politics. The explanation was the same. Neither he nor Marx had any kind of Jewish education. But both had rabbinical forebears, and their genetic inheritance meant they tended unconsciously to adopt the central procedure of rabbinical scholarship, to proceed by commentary on the interpretative work of predecessors. Neither could write from scratch, ab initio. Just as *Das Kapital* is an enormous series of reactions (or comments on) earlier economic writers, so the *Tractatus* is a jerky process of commentary on the logical methods the author derived from Frege and which he found being taught at Cambridge.

All the same, it is highly original, perceptive, stimulating and, above all, exciting. Wittgenstein had the gift Socrates had, to get young people worked up about the way they thought. Few who read the *Tractatus* take it all in. All of any intelligence get bits, and the vast majority get a thrill. This despite, or perhaps because of, an eccentric translation, by C. K. Ogden and F. P. Ramsey, which produced such nuggets as "The world is everything that is the case," an opening sentence translated from "*Die welt ist alles, was der Fall ist*"; and "Logic must take care of itself," derived from "*Die Logik muss für sich selber sorgen.*" Wittgenstein says: "The world is the totality of facts, not of things," and "it breaks down into independent facts which divide the world up." He adds, "Things like this chair or that tree are not independent of their surroundings and so are not facts." Then: "Facts are in logical space and independent of one another and can only be stated or asserted." That was typical of the way that he induced readers to think carefully about the way they formulated propositions. The result was to make non-philosophers want to do philosophy and philosophers to do it differently. One of the attractions of the work is that it is only seventy-five pages and no one can believe, at first, that it cannot be mastered. It is the successful struggle to master bits of it, combined with the unsuccessful attempt to comprehend the whole, that constitutes the challenge-delight and leads people to march into it again and again (as with Proust).

Having published his tract, Wittgenstein took his own advice and abandoned philosophy, at least in theory. His substitute was teaching children in a remote village in upper Austria. This proved most unsatisfactory. At a certain level, he was a brilliant teacher, opening up a fruitful dialogue of minds. But for the business of getting facts into small heads, he lacked the patience. His impatience could take the form of violence, and Wittgenstein eventually got into serious trouble for striking a small girl. As casual corporal punishment was universal in Austrian schools at the time, his behavior must have been excessive. It is probable that the local people were afraid of him anyway, judging his eccentricities as evidence of evil. At all events, the experiment ended in abject and humiliating failure. It was then that he built his sister a house, and it gave him some satisfaction, though no one outside the family ever considered employing him as an architect. By this point he had rendered himself penniless by making over his share of the Wittgenstein fortune to other members of the family. Why did he not give it away to suitable charities? Because choosing the charities, like believing in God, would have raised insuperable propositional problems. He took the quick and easy way out.

Naturally, sacrificing his large private income meant he had to earn his living. In 1929 he returned to Cambridge to resume philosophy. Russell and Moore arranged for him to be awarded a Ph.D. solely on the basis of his *Tractatus,* which was treated as a thesis. The following year they got him elected as a fellow of Trinity College, and he began teaching and lecturing. He was at Cambridge throughout the 1930s except for a year he spent at his Norway hut, when he worked on a project which eventually became his posthumous *Philosophical Investigations*. He also, in 1933–1934, dictated material which later became the *Blue Book,* and in 1934 dictated the *Brown Book*. He was a philosophical faculty probationary lecturer in 1930–1933, an assistant lecturer in 1933–1935, and he continued to lecture, although he had no formal post (a highly irregular arrangement) until 1939, when Moore resigned his chair of philosophy, and he and Russell contrived to get Wittgenstein to take his place. I list his appointments because they raise an important point of current relevance. Russell and Moore were powerful figures in Cambridge, and they repeatedly used this power to push Wittgenstein forward. That was entirely justified, for Wittgenstein was not only a genius (there has never been any dispute about that) but a tutor/lecturer who had an astonishing impact on undergraduates and electrified the entire philosophy school. But it is worth noting that under today's highly complex and rigidly enforced bureaucratic regulations for universities, none of this could have happened. Wittgenstein could not possibly have found employment in the university. He would have been lost to Cambridge, indeed to British academic life.

Wittgenstein, thanks to the patronage of key dons, taught at Cambridge for eight years in the 1930s. When war came, he worked first as a porter at Guy's Hospital in London, then in medical research in Newcastle, where he rendered extremely valuable and much appreciated services. He then returned to his chair of philosophy, but spoke repeatedly of giving it all up. His last lectures were in the Easter term of 1947 and he resigned the professorship in the autumn. His teaching, then, covered a period of ten years, and it was sensationally successful by any standard, but particularly by the three criteria most commonly applied: first, the impact on undergraduates; second, the influence on other teachers; and third, the lasting value of the material taught. He was one of the most original teachers who has ever lived, both in manner and content. There were six main reasons why his teaching had such an emphatic success. First, it was surrounded in secrecy, thus casting a gnostic glow over the activity. From the start, the number of Cambridge people, both students and dons, by no means all philosophers, who wished to attend his séances (for want of a better word) was greater than could be accommodated easily, and far greater than he himself desired since he believed comparatively few would benefit. Hence his "lectures" were not announced in the usual way in the *Cambridge University Recorder.* Instead, certain dons were allowed to decide which pupils would benefit, and they were informed of times and places. This leads to the second point: selectivity. Those who attended were aware they were an elite, a chosen group considered worthy of receiving the special knowledge Wittgenstein dispensed. This laid the foundation of the discipleship—one might even say apostleship—which became such a feature of the teaching. Third, the lectures were essentially spontaneous. Even though he dictated the *Blue* and *Brown* books, he normally spoke not from notes but from impulse. He appeared to be thinking on his feet, to be creating as the lecture went on, to be formulating an entirely new philosophical system, or rather a way of thinking, from an impulsive, intuitive and highly emotional process of unrehearsed cerebration. What made this even more memorable was that Wittgenstein often strictly forbade his auditors to take notes:

> If you write these spontaneous remarks down, some day someone may publish them as my considered opinions. I don't want that done. For I am talking now freely as my ideas come. But all this will need a lot more thought and better expression.

No one was more profuse in sharing his ideas or more proprietorial in forbidding their unauthorized dissemination. Whenever other people attempted to present his views in print, he always set up a caterwauling of rage. Those who did not acknowledge their debt he accused of plagiarism. Those who did were denounced for misrepresentation. Nevertheless the ban on note taking was not effective. Pupils took notes surreptitiously or from memory later in the day. That is how, for instance, his *Lectures and Conversations on Aesthetics, Psychology and Religious Belief* came to be published.

The fourth reason for success was that the lectures were interactive. The audience was not merely invited but forced to participate (one reason numbers had to be limited). Dialogue developed. He gave students the impression that he was not merely creating before their eyes, but involving them in the creative process. This was heady, unforgettable stuff, and dons who attended were often mesmerized too. Fifth, the lectures were highly demanding. The sessions took not the customary one hour, but two, and like everything else connected with him, were extremely exhausting. People came out "drained"—but also exhilarated. Wittgenstein also insisted that everyone who attended should "sign up" for the entire six- or eight-week course, and should stick to the agreement. He felt this was essential to the continuing process of reciprocating creativity and dialogue, which was the essence of his teaching. Hence, and sixth, those who survived an entire Wittgenstein term felt like war veterans, experienced, battle hardened, honored, decorated—different. They had been initiated into philosophy at the very frontier of discovery. They became members of an expanding club, and that club still exists, in shadowy form, even today. And its Perpetual President is a hero.

Evaluating what they learned—or were taught—is much more difficult than describing the club. The *Tractatus* sprang from Wittgenstein's experience as an engineer, as we have seen. That line of inquiry was submerged in the First World War, and it is amazing, looking back at it from the rest of his career, that the *Tractatus* ever reached written form of any kind. His teaching in the 1930s and 1940s sprang from quite a different background, his experience with small children. His failure in this endeavor to get "facts" into the minds of ten-year-olds led him to believe a better way to do it would be to play games with them, and to use images. This might not have worked with children, but the "language-game" approach he developed served well with students of university age. I suspect he got the idea for the actual language games he devised from reading bits of St. Augustine's *Confessions*. The saint liked to ask questions such as: "Where does the present go when it becomes the past?" etc. Part one of the *Brown Book* contains seventy-two language-game exercises, such as "Imagine a people whose language does not permit them to speak such sentences as 'The book is in the drawer,' or 'Water is in the glass.' They say, rather, 'The book can be taken out of the drawer,' and 'The water can be taken out of the glass.'" Another exercise is "If an animal could read, how would you teach it to become a reading-machine?" The essence of a language game is that this particular philosophical exercise is cut off from the general stream of thought and proceeds according to its own specific rules, which operate as arbitrarily as the rules of any other game, by the consent of the players. The *Oxford English Dictionary* defines it as "a speech-activity or limited system of communication and action, complete in itself, which may or may not form a part of our existing use of language." He himself defined it as "ways of using signs simpler than those in which we use the signs of our highly complicated everyday language. Language games are the forms of language with which a child begins to make use of words." Using such games Wittgenstein established a tighter control of teaching events, and pupils, because he could determine the rules in ways impossible when using math or logic. Such games inevitably reinforced the gnostic element too.

Between the wars the *Tractatus* was widely read on both sides of the Atlantic, as well as in Vienna, where Wittgenstein was a hero-magus. Toward the end of the 1930s, he was becoming famous in English academia, even in Oxford, which traditionally held itself aloft from Cambridge innovations, especially in philosophy. His disappearance during the war whetted appetites, and by the end of 1945, his return to the lecture hall or his presence at any gathering assumed the nature of an epiphany. A Wittgenstein mythology began to adhere to routine anecdotes about his eccentric and inexplicable behavior. Four characteristics attracted particular attention. Foremost was his austerity. He wore no tie and his clothing was of the simplest, bought in workmen's shops. He had simple canvas chairs, or deck chairs, in his rooms. His dislike of any kind of formality became acute. Second, however, he remained magisterial. He spoke with authority (like Christ, as was often pointed out). He could rebuke with crushing power. His friendships, though often powerfully vibrant, living organisms, were scarred with quarrel wounds and often ended in death. Wittgenstein invited people into his life, and then expelled them. Third, Wittgenstein increasingly came to regard academic life as debilitating; it "lacked oxygen," as he put it. Sooner or later he counseled nearly everyone, dons and undergraduates alike, to quit it. Fourth, he developed a detestation of science, or rather, of scientism. As he put it in the *Blue Book,* great damage was done when philosophers "see the method of science before their eyes and are irresistibly tempted to ask and answer questions in the way that science does." He thought scientific method particularly inappropriate in aesthetics and religion. He had harsh things to say about works of popular science, such as James Jeans's *The Mysterious Universe,* which he categorized as "idol worship," the idol being "science and the scientist." What he would have said about the present university climate of scientism, promoted by TV dons such as Richard Dawkins, makes the mind reel. In this respect, as in others, he stood for the old high culture of taste and discernment which had been central to European civilization for a thousand years, and was now threatened by a horrific new kind of materialism.

Socially, he steered an uneasy path between Bloomsbury, to which Russell and Moore had introduced him, and outsiders like F. R. Leavis, who hated Bloomsbury and all it stood for. Keynes tried to patronize him by benevolence, and was told: "Keynes, give up economics!" (It may also be true that he told Virginia Woolf: "Woolf, give up writing novels!") Leavis was one of those with whom he struck up an edgy friendship, punctuated by misunderstandings. They went for long walks together, and sometimes paddled the river in a canoe. After Wittgenstein died, Leavis wrote an account of their friendship. It is written in clumsy Leavisite prose and is in places opaque to the point of incomprehensibility, but it does give, in its own way, a vivid impression of the man: authoritative and sometimes authoritarian, but also impulsive, generous, apologetic; and always fascinating. Inevitably he said: "Leavis, give up literary criticism!"

Wittgenstein's increasingly heroic status among many followers and admirers was enhanced by the mythological anecdotage which gathered around him. One famous or notorious incident took place on Friday, October 25, 1946, at a meeting of the Cambridge Moral Science Club in King's College. The speaker was Dr. Karl Popper, the author of what was becoming a key libertarian text, *The Open Society and Its Enemies*. It is hard to think of two more different approaches to philosophy than those of Wittgenstein and Popper. Not that Wittgenstein was anti-libertarian; far from it. But he did not believe philosophy had anything to do with such things or their antonyms. It was not about *what* to think but *how* to think. The two men quickly exchanged heated words, and Wittgenstein, in his nervousness, grasped and rattled a poker in the grate. When he challenged Popper to give an example of a moral rule, he was answered, "Not to threaten visiting lecturers with pokers." At that

Wittgenstein vanished into the night. But this is Popper's version of what took place. There are several others, irreconcilable on salient points. An entire book has been written about those ten minutes, and those who read it can make up their own minds about the philosophical significance—if any—of the events. Popper could be quite as difficult as Wittgenstein. A quarter century later, having received an enthusiastic letter from Popper about one of my books, *Modern Times,* I invited him to my house in Iver (he was then living a few miles away). He said: "Can you give me an *absolute guarantee* that no one whatsoever has smoked in your dining room for at least six weeks?" "No, I am afraid I cannot." "Then I shan't come."

There was a comparable incident at Oxford, in May 1947, the evening already described, when I caught my only glimpse of Wittgenstein. The meeting, in which he replied to a paper on Descartes's "*Cogito, ergo sum,*" took place in a dingy lecture hall. Mary Warnock noted in her diary: "Practically every philosopher I'd ever seen was there." Wittgenstein characteristically declined to address the subject officially under discussion, grinding instead an axe of his own. Oxford's orthodoxy, in the person of the elder Professor Joseph Pritchard, constantly interrupted him and was very rude. Wittgenstein was rude back, and it is a melancholy fact that Pritchard died a week later. Isaiah Berlin, one of those present, told me: "An execution. I would not have missed it for worlds."

That autumn Wittgenstein gave up his chair, and went to live, remotely, in Ireland. He completed his *Philosophical Investigations,* but then became ill with what was then diagnosed as incurable cancer. In his last months he lived in the Cambridge house of his devoted doctor. He died on April 29, 1951. With the death of Wittgenstein, the cult of his philosophy could begin, and his personality quickly changed from eccentric to hero. Every aspect of his life and thought has since been minutely examined in books, and those who knew, or met him, have had their memories ransacked by eager acolytes. The countless books and essays on him have generated more incense than light, and even the first-class biography of him by Ray Monk has obscure corners. There are, and will remain, important lacunae in our knowledge of this strange genius, and eager researchers and admirers will be poking over the evidence, often fragmentary and disputed, for generations to come. But that is the nature of hero worship. Meanwhile, his memory draws clever young people to the art of philosophy, and that is the practical incense which was sweet to his nostrils, albeit it provoked in the end the imperious admonition: "Stop doing philosophy!"

10

THE HEROISM OF THE HOSTESS: LADY PAMELA BERRY ET AL.

Women have had few opportunities to play heroic roles in a continual sense, as warriors and leaders have. And certain specialist niches in society which they have made their own tend to be overlooked as theaters of heroism. One such is party giving. I see the eyebrows shoot up: what is heroic about being a hostess? I quote the answer given by one of them; try it and see.

The idea of a woman being the convenor of a symposium, or any other gathering, for the purpose of discussion or even mere conversation and entertainment would have been unthinkable in antiquity. Indeed, women were rarely present, except as dancers or handmaidens—or prostitutes. For a woman to be prominent at a feast was an ill omen. The entertainments which Cleopatra staged for Antony in 33–32 BC were an aspect of the decadence which led to his downfall, and the presence of Herodias and her daughter Salome at the feast of Herod Antipas in AD 29 led to the disgraceful murder of St. John the Baptist. In medieval times, kings and lords gave feasts, and it is not until the seventeenth century, beginning in France, that we hear of salons conducted by women, the prototype being the one held in her Paris home, 1618–1650, by Catherine de Vivonne, Marquise de Rambouillet. There, Malherbe and de La Rochefoucauld, Bossuet and Corneille laid down the law on manners and spelling, promoted periphrasis and *préciosité* and waged war against Boileau, Racine and Molière. There followed in due course the famous salons of Marie du Deffand, and her breakaway rival Mademoiselle de Lespinasse, competition being of the essence of *salonières* in their heyday. Bourgeois salons, inaugurated by Anne-Marie Cornuel, competed with the aristocratic ones. Madame Roland used hers to create a political party, the Girondins, and Madame de Staël ran an anti-regime salon against Bonaparte.

Party-political salons were still standard in Paris up to the Second World War. Madame de Portes, who ran one tinged with the Far Right, became the mistress of Paul Reynaud, prime minister of France at the time of the military collapse in 1940. She fought for Reynaud's soul on behalf of the Nazis, Winston Churchill tugging the other way. The tortured man resigned, then fled with her from the advancing German armies in a car piled with luggage. There was an accident, and an enormous, heavy suitcase shot forward from the backseat and neatly took off her head.

A decade later, in the early 1950s when I lived in Paris, I used to see Reynaud swimming, almost daily, in the Piscine Déligny, which was in the Seine, conveniently near the Assemblée Nationale, in which he still sat. He always wore a woman's bathing cap, which had belonged to the decapitated lady. By that time party salons had ceased to exist, though there was a political one run by an American millionaire who had inherited a mining fortune. Of those I attended, one was run by Marie-Louise Bousquet, for smart society people (she was connected to the fashion industry) and one was run for literature and the arts by Natalie Barney, another American expatriate. This was long established, and had been attended by Proust. Such salons were held regularly during the season, on a special day each week, and once you had been properly introduced, you could come when you liked. The company was all, for the fare was unappetizing: sticky biscuits, sweet vermouth, perhaps a glass of champagne.

The French salon system never took root in London. In Dr. Johnson's day, Mrs. Montague set one up, giving rise to his enigmatic remark: "One would prefer to drop Mrs. Montague, than to be dropped by her." But the English never quite liked the *regularity* of the salon. Charles Lamb had his "Wednesdays" but found it irksome, and spaced them out to once a month. Hostesses preferred each event to be sui generis, especially country-house parties, the preferred English form of entertainment, which by their nature could not be regular.

Of the famous hostesses, six stand out. First comes Lady Holland, who entertained at Holland House, the remnants of which (the house as a whole was destroyed by Nazi bombers) is now in a fragment of park in built-up Kensington but was then open country with a large farm attached. She invited you to dinner and to "bring a nightcap" if you wished, since there were dozens of bedrooms, and getting back to central London late at night could be tiresome. Apart from herself, the dinners were men only, for she was a divorced woman: ladies in polite society would not receive her, as a rule, or enter her house. But men, especially Whigs, radicals and writers, were delighted to come, for she kept a fine table and was clever at getting people to talk. The Tories often expressed regret that they had no comparable gathering place on their side. William Pitt had no wife, having turned down (or rather fled) Madame de Staël. Lady Peel, who was beautiful and delightful, and had a suitably rich husband with a fine house (and art collection) in Westminster, on the bend of the river, could have emulated Lady Holland but was too shy. Lady Palmerston was certainly not shy, and entertained often at their big, sprawling house at the bottom of Piccadilly (later the Naval and Military Club and known as the Inn & Out). These "drums," as they were called, took place several times during each sitting of Parliament, and a young man could get in without an invitation if he was handsome, well dressed and had the nerve to present himself. These Palmerstonian drums were described by Thackeray, Trollope and Bulwer-Lytton, though not by Dickens, who would not go, or Wilkie Collins, ditto, because he was not allowed to bring either of his two regular mistresses. Big political receptions of this type were held regularly by Lady Londonderry in the between-the-wars period, at Londonderry House in Park Lane. She liked to receive the guests at the top of the first flight of the main staircase, with the prime minister of the day at her side—first Stanley Baldwin, then Ramsay MacDonald, then Baldwin again. But when Baldwin was succeeded by the moth-eaten Chamberlain, she lost heart and stopped giving her drums (called, by then, "bashes"). Londonderry House, the last aristocratic mansion, on Park Lane, on which Evelyn Waugh modeled Marchmain House in *Brideshead Revisited,* was hired out for wedding receptions and the like in the postwar years, and then ignominiously pulled down.

Lady Londonderry, however, did not give regular parties throughout the two seasons (May to mid-July and October to mid-December). That was the self-allotted job of Lady Colefax and Lady Cunard. Both specialized in lunch parties for ten or a dozen people, but they gave dinner parties and theater supper parties too. Sybil Colefax reckoned to give two lunches and one dinner a week; Emerald Cunard sometimes as many as three lunches and two dinners. Their guests were available royalty, headed by the Prince of Wales; dukes such as "Bendor" Westminster; the more *sortable* cabinet ministers; visiting celebrities; newspaper proprietors like Beaverbrook, Rothermere, Luce and Hulton; and occasional exotics like Josephine Baker, Paul Robeson and the singer-pianist "Hutch." After T. S. Eliot achieved fame with *The Waste Land* in 1923, he was "taken up" by both these competitive hostesses, as was Evelyn Waugh when *Vile Bodies* became a best seller in 1930, and W. H. Auden in the mid-1930s briefly, until his disgusting personal habits caused him to be "dropped." It was as easy to be dropped as to be taken up. Lady Cunard, who had been pro-Edward VIII during the abdication crisis, was punished by the new queen, who let it be known that those who attended her parties would not be welcome in royal circles. Thereafter Cunard slowly fizzled out. Of course both women were hit by the war. Colefax became poorer, and had to change for what she called her "ordinaires," now held at the Dorchester, which possessed the safest basement bomb shelter in the West End. The peace, which brought the "labour occupation" and an intensification of rationing, was the end of their era.

It had already ended for the most important hostess of that generation, Lady Ottoline Morrell, who died in 1938. At the age of seven, she was granted the precedence of a duke's daughter when her half-brother became Duke of Portland (1880), and, after her husband Philip Morrell became a liberal MP in 1906, she began to entertain in London. From 1913 to 1924, the Morrells lived at Garsington Manor in Oxfordshire, and there she held the most famous literary and artistic house parties in English history. Among her guests were Bertrand Russell (for a time also her lover until she found his chronic bad breath insupportable), Augustus John, D. H. Lawrence, W. B. Yeats, Walter de la Mare, Aldous Huxley, T. S. Eliot, Siegfried Sassoon, Henry Lamb, Lytton Strachey and Virginia Woolf. The house was beautiful but also cozy, the food delicious, the garden delightful with many intimate nooks and crannies, and the conversation at a high level but also (at times) thrilling and alarming. Lady Ottoline had a notable gift for spotting and encouraging talent, and promoting it publicly. Lawrence and Huxley in particular owed a great deal to her. The atmosphere at Garsington was unique and entirely her creation, and she herself, as Lord David Cecil wrote, central to it: "Her own personality was, in its way, a considerable work of art, expressing alike in her conversation, her dress, and the decoration of her houses, a fantastic individual and creative imagination."

It was the story of Ottoline Morrell which first made me see the hostess as a potentially heroic figure. She and her husband were not rich: more than half their income went on entertaining. She was a hard worker and devoted immense amounts of energy, both physical and nervous, to presenting Garsington as a place where writers and artists could be at their creative best. The tiniest details were important to her in making her guests feel loved: timing and serving of meals (especially breakfast); masses of comfortable furniture and well-chosen books in the bedrooms, and perfect bedside lighting; the selection of fellow guests to make a creative mixture; and the wide choice of activities and expeditions to keep them occupied when not writing or talking. This success as a hostess of difficult artistic people, testified to by all, was due to her willingness to take infinite pains and never to relax. But it was due chiefly to her warm sympathy for often unhappy or depressed people, her ability to enter into their troubles and give wise counsel and her capacity to encourage. She was a good woman, and one hopes she got her reward in heaven, for it was certainly not forthcoming on earth. The literati, not least those who enjoyed her (unreciprocated) hospitality most often, gossiped spitefully behind her back. Russell was disloyal, Lytton Strachey his customary self—critical, mean minded and inaccurate—Lawrence lampooned her in *Sons and Lovers,* and Huxley presented a parody of Garsington in *Crome Yellow*. Lady Ottoline and her careful, anxious entertaining were used as a pretext for working off grudges against the upper classes, the rich, the pretentious, by the clever, grubby intellectuals she fed, housed and befriended. Nothing new in this, of course: Rousseau had done exactly the same in the eighteenth century, and Wagner in the nineteenth. But she took it hard, being sensitive and in many ways vulnerable—it was indeed her sensitivity and consciousness of how easy it was to be wounded that made her so effective in her chosen role.

Ingratitude is something all society hostesses receive and become habituated to. "You have to do this thing for its own sake and because you like it. There's no other recompense." The speaker was Lady Pamela Berry (later Lady Hartwell), and she said something like this to me several times during our friendship, which began in 1956 and lasted till her death in 1982. She is often described as "the last of the political hostesses," and the title is just. But it has to be understood that she was a hostess as a *pis aller*. She really wanted to be a politician, but felt debarred by her sex. Not that she disliked being a woman. On the contrary, she delighted in femininity, in clothes, makeup, fashion, hairstyles, on being inconsequential, illogical, impetuous, passionate, changeable, perverse, on being able to say, "Oh, but I can't be expected to know anything about *that*," on being outrageously womanly and disgracefully, irresponsibly feminine, on being girlish. She carried around with her a personal space in which the scents of a Salome and the guile of a Cleopatra mingled with the restless provocation of Récamier.

Nevertheless, Pamela gave an overwhelming impression of missing the opportunities of a man. Her life was, in one sense, dominated by the memory of her father, F. E. Smith, first Earl of Birkenhead. Now he was a hero for all seasons, though not for all people. He was a self-made man, and exulted in the fact. When awarded his coat of arms, he chose as his motto *Faber fortuna mea*—maker of my own fortune. He came from Birkenhead, a characterless town on the other side of the Mersey from Liverpool. Smith liked to exaggerate the lowliness of his origins and tell stories, especially later in his life and in his cups, of childhood hardship. In fact, his father was a solid middle-class citizen of the town, who rose to be its mayor, and Smith got a good education at Birkenhead School before going to Oxford, where he took first-class honors in jurisprudence. His college, Wadham, then small and obscure, was just on the eve of its period of greatness. Among Smith's contemporaries were John Simon, another future lord chancellor, and C. B. Fry, the famous cricketer. The last left a portrait of the young F. E.: "A long, lean brown face and an impudent nose. Very remarkable eyes. They were the colour of a peat pool on Dartmoor, full of light and fringed by luxuriant silky eyelashes." He was enormously good looking, tall and athletic, a tremendous competitor in any contest going, from boxing to rugby and, as soon as he could afford it, a first-class golfer, horseman and rider to hounds. But, "the lips were slightly ajar, as if about to close on a cigar, and shaped to violence or disdain." He was hair-raisingly articulate, scintillating and reckless in what he said and, to quote the Tory leader Bonar Law, "It would be easier for him to keep a live coal in his mouth than a witty saying."

In London, where he entered the bar, Smith quickly acquired the mannerisms, accent, self-confidence and arrogance of the Tory aristocrats. He was also a sensationally hard (as well as a quick) worker and made rapid progress at the bar, soon earning over £6,000 a year. He performed particular services for the Liverpool shipping magnates and the great local soap tycoon William Lever (afterward Lord Leverhulme), who was made the target of the Northcliffe press. Lever wanted to sue but a well-known K.C. told him his case was not strong enough. Dissatisfied, Lever asked F. E. for his opinion. F. E. hurried to London, to the Savoy Hotel, where he found a stack of papers four feet high waiting for him, and a note saying his opinion was required by nine AM the next morning. He ordered two dozen oysters and a bottle of champagne and sat up all night reading. Promptly at nine AM he gave his opinion, one of the shortest on record: "There is no answer to this action for libel, and the damages must be enormous." They were: £50,000.

This notoriously laconic opinion was given in 1906, the same year Smith, aged thirty-four, entered Parliament for Walton, Liverpool. The famous 1906 election was among the most disastrous in Conservative history. From having a majority of 134, they found themselves with barely 150 seats, while the liberal majority was 356. But this nadir of Tory fortunes gave Smith a matchless opportunity to make his name at a stroke by restoring Tory morale in his maiden speech, delivered at ten PM on March 11. Traditionally, a maiden speaker "craved the indulgence" of the House in hearing him, and repaid it by keeping his material uncontroversial. Disraeli ignored this tradition and was howled down. Smith likewise craved no indulgences and made his speech as offensive to the government benches as he possibly could. He prepared it with great care, spiced it with witticisms and superb jokes, learned it by heart and delivered it with stunning confidence, keeping a straight face throughout. When F. E. first stood up, the Tory benches, meagerly attended, were a picture of misery and dejection. When he sat down, fifty minutes later, the House was packed, for word had gotten around that a great event was taking place, and the Tory benches were a shouting, foot-stamping mass of jubilation and excitement. It was, without question, the most famous maiden speech in history, quite unprecedented and never equaled since. F. E., who had to take the midnight train north, woke up in Chester, like Byron a century before, "to find himself famous."

Thereafter he flourished, a leading buccaneer in the last great age of British political privateering, on the high seas of the Commons. His comrades in adventure were nominally his opponents, Lloyd George and Churchill, united with Smith in their oratorical brilliance and boundless ambition. Smith became attorney-general in the wartime coalition, and when Lloyd George remade his government after the landslide coalition victory of 1918, he raised F. E. to the woolsack as lord chancellor and, in due course, gave him an earldom. "LG," Birkenhead and Churchill were the three titans of the postwar coalition, riding high over the mediocrities who, in their view, constituted the rest of the House of Commons, and then coming crashing down, in the autumn of 1922, when the mediocrities rose in revolt against LG's high-handed ways and threw his government out. LG never held office again, Churchill was forced to go cap in hand and rejoin the Tory Party, which he had left in disgust twenty years before, and Birkenhead became a spent force, drinking heavily and getting himself into debt.

Nevertheless, the earl had a shot or two left in his arsenal. Two days after the fall of the coalition, he had been elected lord rector of Glasgow University by the students, the losers being his old rival John Simon and H. G. Wells. On November 7, 1923, he gave his rectorial address on the subject of "Idealism in International Politics." The speech was, in its own way, as sensational as his maiden, and required the same kind of courage. At that time, the League of Nations was regarded by all, or almost all, as the ultimate solution to all international problems, and young people were expected to pay it unqualified obeisance. The era of power politics, they were told, was over for good, and perpetual peace had taken its place. The young must prepare themselves for a lifetime of idealism and do-gooding, and expect a halo in due course. All this, said Birkenhead, was absolute nonsense. The world would continue to be a rough place, with force the ultimate arbiter and war always just around the corner. He conceded that there might be "a modest area within which the League of Nations may make useful contributions to the harmony of the world [but] the larger claims made on its behalf always seemed to me to be frankly fantastic. Its framers forgot human nature as absurdly as they neglected history." As for youth, it would do well to note the facts of life. "The motive of self-interest not only is, but must be, and ought to be, the mainspring of human conduct." Only "the desire of self-advancement" was an adequate incentive "to the labour required to push the world forward." Fortunately, he concluded, "The world continues to offer glittering prizes to those who have stout hearts and sharp swords."

There is no doubt that what Birkenhead said about the league and world peace was true, prophetically so, and was to be proved so in the next fifteen years—and the advice he gave to youth was good, as the Glasgow students themselves recognized, applauding the speech. But their elders were shocked, indeed outraged. Birkenhead was almost universally denounced as a disgraceful example of cynicism, gross materialism and worldiness. It was the time when the extreme liberalism and pacifism of Bloomsbury was taking over and Bertrand Russell, Lytton Strachey and John Maynard Keynes were setting the moral (or, some would say, the immoral) tone of society. Birkenhead went head-on against this trend, and though he was eventually proved right, at the time he was treated as a leper, especially by the bishops and other clergy and by those in society usually described as "right thinking."

The criticism of Birkenhead was all the more fierce in that, by the mid-1920s, he was clearly out of the running for the prime ministership. Baldwin made him secretary of state for India, but he was eventually forced to resign in order to go into the City of London and make money. By this time there were ugly rumors about his drinking and his debts. His son and heir, "Freddie," second Earl of Birkenhead, wrote in his biography of his father:

> At the end of his life, when he had left politics for the City, he had his yacht, six motor cars, only three of which were normally used, three chauffeurs, eight horses with three grooms, a large London house in Grosvenor Square, and a house in Oxfordshire. He refused to attend to his income-tax returns, exercised no control over his agents and gradually acquired an enormous overdraft. This rake's progress was prompted by the same hubris which led to the downfall, in a different way, of Oscar Wilde. When protests were made to him he bought another car or a new motor-launch, and this selfishness and indifference to the interests of his family was undoubtedly the least attractive feature of his character.

These are harsh words for a son to use about his father, and seem all the more repellent in that everything the second earl enjoyed in life, his expensive education at Eton and Oxford, his peerage, his connections and the patronage of the great, were the direct result of his father's hard work and brilliance. He achieved very little, no doubt because he failed to inherit either his father's brains or his judgment, perception and capacity for intense and determined industry.

By contrast, Pamela was her father's child through and through. The youngest of the family, she was also the wildest, the cleverest, the most imaginative, the fiercest and the naughtiest. When the Lloyd Georges and the Birkenheads shared a holiday at Algeciras in January 1923, Pamela, then eight, earned this accolate from LG: "His youngest kiddie—'the Lady Pam'—is a terror—utterly spoilt." It was true. Birkenhead adored Pam, born just before the First World War, in May 1914, and let her do anything she wanted—except go to proper schools.

It was Pam's great grievance that her father's views on women were so old-fashioned that he insisted girls be educated at home. "As a result I am almost totally uneducated, and unfit for any job." This was the only criticism I ever heard her utter of her father, whom she loved and hero-worshipped, and whose memory and reputation she defended with the ferocity of a tigress. She was seventeen when her father died, suddenly, not yet sixty, in 1930. There was a dreadful financial reckoning. The London house, yacht, motor car, horses, had to be sold, servants dismissed, and his widow was able to remain at the Oxfordshire house only through the generosity of friends. One of them, Lord Beaverbrook, who had once described F. E. as "the cleverest man in the kingdom," came to the rescue of the children, making each an allowance of £325 a year. This enabled Pam to have a "season," and in due course she made a splendid marriage to Michael Berry, son of Lord Camrose, who owned the *Daily Telegraph*. She then told Beaverbrook that the allowance should cease, but he would not hear of it.

Michael Berry became editor in chief and, in effect, the boss of the *Telegraph* group of newspapers, and it was as his wife that Pam set up shop as a London hostess. She felt herself (rightly or wrongly) barred from politics by her sex, for she despised the type of woman who in those days (I am talking about the 1940s) got into the House of Commons—they were still excluded from the Lords—and she thought her lack of education disqualified her from any other career. Entertaining the elite, and especially the political and intellectual elite, was thus for her a vicarious way of engaging in public life, and she set about the business with voracious enthusiasm, what became highly professional skill and overwhelming success. For more than thirty years she held the field. None of her rivals even approached her in her ability to gather fascinating people around her table and get them to sparkle. And when she died there was no one to take her place—nor has there been in the quarter century since.

What makes an outstanding hostess? First is intelligence. It is the essential quality because the work is supremely difficult and requires it. The idea that a woman can succeed in it simply by possessing lots of money and social position—even if you throw in charm and good looks—is nonsense. It needs brains. Pam inherited her father's brains and his sense of adventure but she also possessed a prudence he lacked. She had to be as clever (or almost so) as the cleverest of her guests but also disciplined enough to hold her tongue and let them speak. It is a curious fact that, although Pam was talkative (none more so) on the phone or in a tête-à-tête, and sharp, spicy and delightful talk it was, she hardly said a word at her lunches. She set hares running, she occasionally pricked conceit, she might even rebuke, briefly, but her words usually served only to flick forward deftly a faltering conversation—something she rarely needed to do—and as a rule, she remained silent, watching, always ready to come to the rescue of a fragile guest who was being bullied or ragged, but for the most part just laughing, encouraging and applauding. She said to me: "I am at my best when nobody feels I am there." Or again: "Wasn't there someone who referred to the Invisible Hand?" "Yes, Adam Smith in *The Wealth of Nations*." "Oh, goody, darling, you know everything. Well, I am the Invisible Hand. I do the guiding, but nobody sees. Nobody wants to feel the conversation is being 'directed.' And it isn't, actually, is it?" "No, the best talk is entirely spontaneous and off the cuff. Can't bear people who prepare Good Things in advance, or hostesses who prod, prod, prod all the time. A good conversation is what that American general-person, what's his name, yes, Haig, called 'a controlled explosion.' Or a series of controlled explosions, leading to the Big One."

Brains were needed to handle the extraordinary assortment of people who came to her table. Evelyn Waugh was superlative on form, but had to be coaxed out of bottomless misery at times. Dick Crossman required to be stopped from shouting people down, as had Randolph Churchill (but he was often banned). Tommy Balogh might be unwittingly provoked into an uncontrolled explosion of prewar-Budapest rage. Ted Heath, one of Pam's unaccountable favorites, was one of the rudest men in London, not out of malice but by nature, and needed careful *placement*. Laurence Olivier required royal dollops of imperceptible flattery. Some visiting Americans, like Arthur Schlesinger, were up to all the latest London scandals but others—Danny Kaye, Arthur Miller, General Westmoreland, Governor Stevenson and Clare Luce—needed protection from incomprehensible in talk. Tom Driberg was often beastly to servants and required threatening. Duff Cooper, a man of brief, spasmodic outbursts of fury, known as "veiners," might not be on speaking terms with one or more of the other guests. His wife, then his widow, Diana, was angelic but might be tipsy, reminding one of Lamb's description of Coleridge, "an archangel slightly damaged." Princess Margaret was always what Professor Bronowski, a grand master, called "a three-move problem." Graham Greene might be monosyllabic or mute. Ian Fleming itched to get back to Bond or, even more likely, golf. His wife, Annie, was a rival hostess, looking for slips to be exploited. The frogs, as Pam called them, were particularly difficult: André Malraux needing first-rate flattery in faultless French, Nancy Mitford's *colonel* horribly particular about the food, Louise de Vilmorin stirring up jealousies, male and female. Pam had her own list of those she termed prima donnas, marvelous to have if well-behaved but liable to sulks: Anthony Eden, Cyril Connolly, Anthony Crosland, Maurice Bowra.

A hostess is like a theater director with a tricky cast and an improvised play. Or like a zookeeper with animals in unlocked cages. Pam used to say that twelve was the right number for lunch—never more—as that was the maximum number of interesting people she could control. This was, as she intended, intimate entertaining. Her house in Westminster was made for it: chosen because it was in the Division Bell area, convenient for ministers in Whitehall to pop over for lunch or MPs to have dinner during a busy evening's voting. The rooms were small and cozy. She could, if necessary, double the size of the dining room but she liked to keep things small. There was breathtaking attention to detail, born of long experience, excellent taste and a powerful mind. The food was perfect but never excessive or awkward to handle; one was never conscious of the wine flowing, as it appeared in your glass as if by nature; everything was served imperceptibly and with infinite skill, at exactly the right temperature. Nothing distracted from or impinged upon the talk.

Pam was not just a highly intelligent woman. She was also deeply emotional, with feelings that at times overpowered her. She had hot blood. She loved passionately, ruthlessly and, at times, recklessly. She was also a superlative hater. And her hates could be long lasting, indeed perpetual. Once she had decided firmly against a person, she was unremitting in her antagonism and reasoned arguments could not shake her. Her obstinacy was monumental and granitic, her chilliness marmoreal. Once or twice I tried hard and repeatedly to argue her out of a particular antipathy toward someone who remained a friend of mine and whom I knew she had misjudged. But she rarely revised a guilty verdict. "Pam, you have got your Hanging Judge Face on," I would say. This was the other side of her deep and glorious passion for friendship. This too was inherited. Winston Churchill said of F. E., when he died, that their friendship had been unbroken for nearly a quarter of a century, from the day they first met properly and warmed to each other. He said that F. E. was not one of those acquisitive men who piled up estates and possessions, "piles of scrip and bonds and shares," and tangible assets to be tucked away in vaults and strong rooms. Rather, "he banked his treasure in the hearts of his friends." I often think of this noble phrase, and debate how far it applies to people I know. Certainly Pam banked her treasure in her friends' hearts, and expected them in return to bank their treasure in hers. Friendship, close, intimate, firm, indestructible and perpetual friendship, was the guiding principle of her life. Along with her intelligence, it was the reason she was a hostess without rival, for she saw her guests as possible, potential, actual friends. Some turned out duds in this respect, and others even became enemies. But most, in all kinds of different, sometimes very peculiar ways, were friends, or treated as such, with tenderness, love, devotion and the affection which a guest always needs and so seldom receives. She made her guests feel special because she always tried to make them her friends too, as the dwindling band who still remember will testify.

Pam gave many large-scale parties, too, for public occasions. She always had a huge election-night party at the Savoy, down the road from the *Telegraph* fortress. This was the scene alternately of unbecoming exultation or gruesome despondency, guests leaving or arriving all the time into the small hours, having triumphed with obscene relish or having lost their jobs and often their seats too. Pam presided over these long nights of electoral carnage or indecent joy with extraordinary patience and almost sacramental charity. I thought of her, at these times, as a high priestess of England's political faith, stretching out a consoling or restraining hand to vanquished and victor alike. She had the personality and the training, and the wise self-discipline, to do this with elegance and professional aplomb. And of course, as at her lunch parties, every detail was planned and supervised with immense care, so that the stricken were comforted, the exuberant calmed, the worried reassured, the mighty made to feel humble in their hearts. It was tremendously hard work, often emotionally exhausting too, for Pam felt deeply for a politician who had just seen his career tumble down in ruin. Once, after a particularly stressful night, I was up early, and had masses of flowers sent round to Pam's house in Cowley Street. The phone rang: "Do you know, you are the only person who ever sends me flowers after a party? The others always take it for granted." Not true, of course, but Pam sometimes was overcome by that feeling which all hostesses experience. What's the point? Are they worth it? Is it all a waste of time and money and effort, and of love? That is why I see the hostess among the ranks of the heroes and heroines, and usually unsung too. Certainly Pam was in my Valhalla, a valkyrie of the festive table, a Brunhilde of the fête.

11

A GENEROUS HERO AND A HEROIC MONSTER: CHURCHILL AND DE GAULLE

Churchill was the archetypal hero of the twentieth century, and his life was so long, and full, and varied that one has to concentrate on salient points to sharpen the profile. First, health and energy. There were various physical crises in his life, including a near-fatal accident on Park Avenue in 1931 and heart attacks in the war and postwar. But on the whole his health was astonishingly good, as emerges clearly from the book by his personal physician, Lord Moran, *Winston Churchill: The Struggle for Survival* (1966). (This is one of the two really good books about him, the other being *The Winston Churchill As I Knew Him* (1965) by Lady Violet Bonham Carter, dealing with the years 1906–1918).

Why did Churchill enjoy such good health? He took little regular exercise. He ate exactly what he wanted. Most people believed he drank too much, especially of whiskey and brandy. After brandy, said Cyril Asquith, "he looked like a tortoise." But he sipped the brandy very slowly, and he took his whiskey heavily diluted with water. He never gulped liquor. In 1949, aboard Aristotle Onassis's yacht, *Christina,* he boasted of his alcohol consumption, and said:

"If all the whisky and brandy I have drunk in my life was added up, it would fill this state-room to overflowing." Professor Lindemann, Lord Cherwell, his scientific adviser, was present, and doubted Churchill's assertion. So he was given his master's daily intake, the measurements of the stateroom (which, like everything else on the yacht was of great size) and told to get out his slide rule. Cherwell calculated that the spirits Churchill had drunk in his life would fill the room only up to a depth of five inches. Churchill was mortified. After he died, an autopsy was carried out and various bodily organs were examined. His liver was found to be in a perfect state, "like a five-year-old."

Churchill was both a very active and a very inactive man. He spent an extraordinary amount of his time in bed. This was not sloth but method. The first time I met Churchill, in the autumn of 1946, when I was still a schoolboy about to go up to Oxford, I asked him this question: "Mr. Churchill, sir, to what do you attribute your success in life?" He replied, instantly: "Conservation of energy. Never stand up when you can sit down, and never sit down when you can lie down." Curiously enough, exactly the same principle was laid down by Mae West: "I never walk when I can sit, or sit when I can recline." Churchill followed his axiom. If possible he spent the morning in bed, like a duchess in the good old days. But he was not inactive. He read his letters, and dictated answers, went through the newspapers, and did a good deal of telephoning. He also received visitors. Once seated, especially in an armchair, he was reluctant to rise. That aroused the indignation of President Theodore Roosevelt, who said (1904): "That young man is not a gentleman. He does not rise to his feet when a lady enters the room."

Churchill's record of activity is unlikely ever to be equaled. In addition to being prime minister for a total of nearly nine years (for much of which he was also minister of defense), he was twice first lord of the admiralty, secretary of state for war, and secretary of state for the colonies. As chancellor of the exchequer, he carried into law five consecutive budgets, a record equaled only by Walpole, Pitt, Peel and Gladstone, and unlikely ever to be broken. He published over 5 million words. His six volumes on the Second World War alone total 2,050,000 words (including appendices and index). For purposes of comparison, Gibbon's *Decline and Fall of the Roman Empire* is about 1,100,000 words. Churchill built walls, made lakes, painted over a thousand canvases, made thousands of speeches all over the world, some of the finest ever delivered, and mastered the art of broadcasting in a way never surpassed. He is the only member of the House of Commons who, after sitting in the cabinet, commanded an infantry battalion in action. He also survived a number of personal and political crises of great ferocity, notably when he was blamed for the Dardanelles disaster, and when he made a fool of himself during the abdication crisis in 1936. On the second occasion, wrote Lord Winterton, the longest-serving MP, Churchill was the villain of an ugly and humiliating scene, "one of the angriest manifestations I have ever heard directed against any man in the House of Commons." At one time or another, Churchill was viciously attacked from every quarter of the political spectrum. He fought or threatened more libel actions than any other MP in the twentieth century. In 1923 he had the poet Lord Alfred Douglas sent to prison for six months for criminal libel, and threatened to do the same to Clifford Sharp, editor of the *New Statesman,* in 1926. The Tories elected him their leader in the autumn of 1940 only with the greatest reluctance, and many hated him in their hearts, as he did them. The period of his life to which he turned back with the greatest satisfaction were the years 1906 to 1912, when he and Lloyd George were laying the foundations of the British welfare state. Right at the end of his life, he had a curious conversation with a friend of mine, "Curly" Mallalieu, MP. Churchill, nearly ninety and frail, was allowed to use a facility called "the House of Lords Lift" to save him going up the stairs, in which Parliament abounds. Other MPs suffering from temporary debilities were also permitted to use it, and Mallalieu had hurt himself playing football. One day, he found Churchill already in the lift and eyeing him narrowly: "Who are you?" "I'm Bill Mallalieu, sir, MP for Huddersfield." "What party?" "Labour, sir." "Ah. I'm a liberal. *Always have been.*"

Next to energy, Churchill had intelligence. This aspect of his armory for life has generally been underestimated, not least by his parents. It recalls Lady Mornington's foolish dismissal of her son, the future Duke of Wellington (another highly intelligent man) as "food for powder, nothing more." Lord Randolph Churchill and Lady Randolph, disappointed by Churchill's poor academic performance at Harrow, thought him fit only for the army: and, in the army, fit only for the cavalry, which demanded fewer brains than the foot guards or the light infantry. Later in life, he was sneered at by intellectuals such as A. J. Balfour, Harold Laski and R. H. S. Crossman, mainly because he did not use their academic vernacular. But Churchill abounded in natural intelligence, and though almost entirely self-educated (chiefly as a subaltern in India), he was capable of arguing through all kinds of problems. It was the philosopher Karl Popper who first drew attention to the sharpness of Churchill's analytical intellect. He quotes a passage from *My Early Life,* Churchill's fascinating volume of autobiography, first published in 1930:

> Some of my cousins who had the great advantage of university education used to tease me with arguments to prove that nothing has any existence except what we think of it. I always rested upon the following argument which I devised for myself many years ago. Here is this great sun standing apparently on no better foundation than our physical senses. But happily there is a method, apart altogether from our physical senses, of testing the reality of the sun. Astronomers predict by mathematics and pure reason that a black spot will pass across the sun on a certain day. You look, and your sense of sight immediately tells you that their calculations are vindicated. *We have taken what is called in military map-making a "cross-bearing." We have got independent testimony to the reality of the sun. When my metaphysical friends tell me that the data on which the astronomers make their*

> *calculations was necessarily obtained originally through the evidence of their senses, I say "No." They might, in theory at any rate, be obtained by automatic calculating-machines set in motion by the light falling upon them without admixture of the human senses at any stage.* I reaffirm with emphasis that the sun is real, and that it is hot—in fact as hot as Hell, and that if the metaphysicians doubt it they should go there and see.

Popper's comment is as follows:

> Churchill's argument, especially the important passage I have put in italics, not only is a valid criticism of the idealistic and subjectivist arguments, but is the philosophically soundest and most ingenious argument against subjectivist epistemology that I know . . . The argument is highly original; first published in 1930 it is one of the earliest philosophical arguments making use of the possibility of automatic observatories and calculating machines, programmed by Newtonian theory.

It is possible to find many other examples of Churchill's intellectual ingenuity dotted about his writings, and not, as in this case, acknowledged by a great philosopher. The truth is, Churchill had a fine mind, and the fact that it was an undisciplined and uneducated mind sometimes worked to his advantage.

Churchill prided himself on his liberalism and his (by the standards of 1906) progressive notions. He was also a skeptic. It is a curious fact, which some of his warmest admirers find hard to accept, that he had little religious belief. Indeed he regarded orthodox religion as a sham. When he was educating himself in India in his early twenties, he read Winwood Reade's powerful but crude anti-religious tract, *The Martyrdom of Man,* written a generation before, in 1872. It made a huge impression on him, which was never quite effaced. He wrote to his mother after finishing the book:

> One of these days the cold bright light of science and reason will shine through the cathedral windows and we shall go out into the fields to seek God for ourselves. The great

> laws of Nature will be understood—our destiny and our past will be clear. We shall then be able to dispense with the religious toys that have agreeably fostered the development of mankind.

Churchill shared the belief, common among nineteenth-century intellectuals, that organized religion was "the opiate of the people." It was the enemy of progress and of worldly success. He said he believed

> that people who think much of the next world rarely prosper in this; that men must use their minds and not kill their doubts by sensuous pleasures; that superstitious faith in nations rarely promotes their industry, that in a phrase, Catholicism—all religions, if you like, but particularly Catholicism—is a delicious narcotic. It may soothe our pains and chase our worries, but it checks our growth and saps our strength.

This was not just a young man's iconoclasm either. When he wrote *My Early Life* he was in his late fifties. He still held that Reade's book was "a concise and well-written universal history of mankind," proving that, in dying, "we simply go out like candles." Churchill had even less respect for the Church of England than he had for Catholicism, regarding it as essentially a social institution, perhaps of some value as such, but not a depository of transcendental truth. There is no evidence he ever developed a belief in life after death. He sometimes found it convenient or amusing or provocative to posit the existence of a divine providence. But he did so not in the deeply serious tone of a Lincoln, whose religious faith or nonfaith was similar, but in the manner of a man sharing a cosmic joke. Thus, when early in 1951, still leader of the opposition, he told a group of Tory MPs in the smoking room: "It is fortunate that this war in Korea has come while Labour is still in power. We had no alternative but to fight, but if I had been Prime Minister, they would have called me a warmonger. As it is, I have not been called upon to take so invidious a step as to send our young men to fight on the other side of the globe. The Old Man has been good to me." Sir Reginald Manningham-Buller (puzzled): "What old man, Sir?" Churchill: "Why, Sir Reginald, *almighty God,* the RULER OF THE UNIVERSE!"

But if Churchill had no religious faith as such, he had a system of pietàs, centering around Britain's unwritten constitution, the Houses of Parliament, and the House of Commons in particular. There lay his household gods, and he honored them for seventy years of public service. He said more than once that "parliamentary democracy is a very unsatisfactory form of government," and it certainly dealt him some painful blows at times, but it was nonetheless "immeasurably superior to any other." He loved power, and sought it greedily always, most anxious to possess it "in all its plenitude" (a favorite phrase) and most reluctant to relinquish it, keeping the unfortunate Anthony Eden waiting for two nail-biting years as his career drew to a close. He said that the Dardanelles campaign was a failure "because I did not possess the power to make it a success," and he applied the lesson in the Second World War, insisting that he be invested with sufficient power to direct the war effectively. As he put it, "I acquired and exercised power in ever-growing measure." There is a vignette of him, face stern and proud, striding up and down the cabinet office, clenching and unclenching his fists, and saying: "I want them to *feel my power.*" And he was referring not to the enemy but to his colleagues.

Yet in the pursuit and enjoyment of power, he was always not merely careful but punctilious in observing the constitutional rules and respecting those persons and institutions charged with upholding them. This to my mind is the quality in Churchill which makes him so quintessentially the democratic hero. His relations with the sovereign were always exemplary. Watching Churchill bow to the newly enthroned Queen Elizabeth II, a young woman less than a third of his age, was a magnificent exercise in constitutional deportment. He bowed not just to a person but to an institution and to a historical process over a millennium old. And he bowed with genuine humility and a kind of childlike love. History, to him, was the great teacher, and he remained under her iron rod all his life. Along with the sovereign, as head of state, he revered the Commons, ultimate source of the power and authority he enjoyed. It had given him the power; it could take that power away unless he exercised it prudently. Hence satisfying the Commons was always at the center of his preoccupations as prime minister. There were all kinds of complaints laid at his door, but never once was he accused of treating the Commons with indifference, let alone contempt. He respected its offices too, above all the speaker, whom he always obeyed, even in moments of great emotional stress.

Churchill was particularly careful during his wartime days of power to observe the rules which governed his relations, as political master, with the service chiefs, as executives. It would not be true to say that he never sought to teach them their business. He had his views on warfare, often right, sometimes wrong, always well informed and invariably argued with force, even genius. But in the end he always submitted to their judgment about how the policies of the war cabinet were to be implemented. He restrained his skepticism or anger or will—it was sometimes a huge effort—but he kept to the rules. And that was one reason the generals and admirals and air marshals, who often found him a trial and a burden, respected his leadership.

The tensions and efforts of Churchill's life were always made more bearable by his love of jokes and laughter. His sense of humor was not ubiquitous—he could sometimes be stony faced when confronted by absurdities others found hilarious. But it popped out again pretty quickly. He was never glum for long. At his first meeting with Violet Bonham Carter, in 1906, she recalled him saying: "We are all worms. But I really think I am a glow worm." The glow came from a sense of fun that was never wholly extinct, and sometimes radiated warmth of a metaphysical kind. His jokes ran the gamut, from horseplay through whimsy and irony to self-deprecation. They were seldom cruel or savage. He found Sir Stafford Cripps a trial, but the worst he ever said of him, as he passed through the central lobby, was: "There, but for the grace of God, goes God." The only political enemy he could not joke about was Aneurin Bevan ("a squalid nuisance"). He invented the cab joke, about the (comparatively) insignificant Clem Attlee: "An empty taxi drew up outside the House of Commons, and Mr Attlee got out." He made one nasty joke about the detested Vic Oliver, the stand-up comic who married Churchill's stagestruck daughter Sarah. When Mussolini's misfortunes became overwhelming, Churchill commented: "Well, at least he had the pleasure of murdering his son-in-law." Churchill bore rancor at the time but seldom afterward. He said himself he was not a good hater: "The moment victory came, I ceased to hate Germany." His behavior toward Chamberlain was noble. "In war, resolution, in defeat, defiance, in victory, magnanimity, in peace, good will." That was the central maxim of his life. The public turned him out once or twice—spectacularly in 1945. But he always accepted the electoral verdict without recrimination. When suddenly turned off in July 1945 after five years of magisterial authority, he was taken aback. His wife said: "Perhaps it is a blessing in disguise." He replied: "It appears to be very effectively disguised."

In fact, Clementine Churchill was right. Not only did dismissal by the electorate save Churchill from making strategic errors of judgment in the immediate postwar period, especially over India, where he would have tried to hang on, but enforced idleness, after his hectic industry during the war years, made it possible for him to get down immediately to the business of telling his side of the wartime story. In this book I have drawn attention to the efforts of some great actors in history to get posterity on their side. Alexander arranged for writers to put his case. Caesar wrote his commentaries on the Gallic and civil wars. But Churchill was by far the most thorough and successful in ensuring that his tale was told. After his electoral defeat Churchill buckled down to the task of his war memoirs straight away, and by the time he returned to power at the end of 1951, the bulk of the work had been done. If he had remained in office, it might never have been done at all.

The work has five remarkable characteristics, all of which were described in detail in a remarkable book by David Reynolds, *In Command of History: Churchill Fighting and Writing in the Second World War* (2005). First, and perhaps most important, the book was a documentary history as well as a personal memoir. Churchill had always been a hoarder of papers (as was George Washington), and from an early age had kept a personal archive, preserving every scrap of importance about himself. He had earlier dealt with the First World War in about 750,000 words, provoking a famous quip from his cabinet colleague (and former political opponent) A. J. (later Earl) Balfour: "Winston has written an enormous book about himself and called it *The World Crisis*." He learned a lot from the experience, especially the need to get possession and make use of official papers. He applied this lesson to the Second World War from the start. It is likely that many of the most important assessments and telegrams Churchill wrote while the war was on were classified by him at the time with a view to future use in his memoirs. When he left Downing Street in summer 1945, he made what has been called a "remarkable bargain" with the then cabinet secretary, Sir Edward Bridges, the custodian of government documents. Under the bargain, a vast quantity of his wartime papers were classified as his personal property, and he was allowed to remove them physically to the archive at his country house, Chartwell, in Kent. The only condition was that their publication had to be approved by the government of the day. This bargain not only made it easy for Churchill to document his work in full but also benefited the Churchill family financially. He donated all these papers to the Chartwell Trust and this, in turn, sold them to Lord Camrose, the owner of the *Daily Telegraph*, as part of a clever legal device to avoid the punitive taxation which would have made the memoirs financially pointless.

The second characteristic was size, which I have already described. Thanks to the documentary element, all the important episodes in the account, and many minor ones, are described in considerable detail, all of which were new at the time of publication. The work, whether serialized or read in book form, was thus fascinating, despite its size. This raises the third point. By any standards, this was one of the most successful books ever written, both financially and in terms of readership. The original book deal of May 1947 (for a projected five volumes) brought the author $2.23 million, which in today's terms might be anything up to $50 million. In addition Churchill got huge sums from Time Life and the *New York Times* for serial rights. The work helped to win him the Nobel Prize for Literature in 1953. This meant Churchill joined the elite group of Britons such as Kipling, Shaw and T. S. Eliot, and which included only one other historian, Theodore Mommsen, the German chronicler of ancient Rome. At the time of the Nobel Prize winning, the *Daily Telegraph,* which serialized the last volume in London, stated that volumes one to five had already sold 6 million copies, and extracts had appeared in fifty newspapers in forty countries, concluding: "No book has ever so swiftly achieved such dissemination." Both British and American publishers made fortunes from the enterprise, as did Churchill's agent and Riviera host, Emery Reeves.

Fourth, the work was both a personal achievement and a team effort. It was based upon what the author called "the three Ds—documents, dictation and drafts." The wartime minutes and telegrams formed the core, but Churchill also dictated a number of passages on key episodes that he remembered with particular intensity. There were also drafts on the background history of events put together by "the Syndicate," the team of research assistants headed by Bill Deakin, Henry Pownall and Gordon Allen. In addition, other experts, industry service chiefs who held important positions in the war, were called in to supply drafts of particular events, not always successfully. Bill Deakin, an academic who was the only professional historian in the Syndicate, played an unusually important role in revising the entire text of the volume, *The Hinge of Fate,* when Churchill lacked the time to do it himself.

The fact that Churchill was not the sole author (though always the final authority on what went in, or not) does not diminish his achievement. The production of the work may be compared to the efforts of a big research group in science directed by a major figure who gets the credit. Denis Kelly, the office manager of the Syndicate, varied the metaphor. Asked if the great man really wrote the book himself, he replied it "was almost as superficial a question as asking a Master Chef, 'Did you cook the whole banquet with your own hands?'"

Fifth, and this links up with the first point, the work had an impact on how we see the history of the war and its aftermath, which for many decades was almost determinative, and still helps to shape the received history. Here the documentation was all important. The British documents to which Churchill alone had full access were permanently locked up (except to certain select official historians) until in 1958 Parliament enacted legislation on access, but even this concession included the "Fifty Year Rule," which meant half a century had to elapse before the public could see a particular document. In 1967 this was reduced to thirty years, but by then Churchill was dead, having got in with his documents well ahead of the pack.

Churchill was also lucky in that, during the seven years it took him to write and publish the work, he had virtually no competition. Hitler, Mussolini and Roosevelt were dead. Stalin wrote no memoirs, thinking official Soviet history would do instead—and he was also in power continuously until his death in 1953. Being in opposition, and moving (all things considered) with all deliberate speed, Churchill published long before the various generals, admirals, air marshals and politicians who might have helped to shape the vision of history. In effect, the period of revisionism did not begin until after his death, and by then many of the verdicts he sought to impose had already become part of the concrete and granite of historical teaching, exceedingly hard to undermine and replace. It is difficult to say how widely the work is still read, half a century after it was published, though as there are still millions of copies of it still on the shelves of public and private collections, one must assume it is still dipped into. But in a sense the question is irrelevant, since the message of the work, and many of its details, have passed into the public historical memory, at least in the English-speaking world, and probably much wider.

To a great extent then the heroic epic which gathered around Churchill was written or inspired by himself. From a study of how his *War Memoirs* were done, including deliberate omissions, suppressions and manipulations, the impression that emerges is that Churchill was a historian of great passion, insight, romantic and almost poetic gifts, huge industry and remarkable power, but also an operator of considerable ruthlessness. In writing his masterpiece describing the course of the greatest of all wars, and his own role in it, he knew very well that he was fighting for his place in history and his heroic status. He fought hard and took no prisoners, and on the whole he won the war of words too, as he had earlier won the war of deeds. And, as he once truly remarked (May 1938), "Words are the only things that last for ever."

Churchill's younger contemporary, Charles de Gaulle (1890–1970), was a national hero too, who saved the country twice, first in 1940 from shame and ignominy, second in 1958 from civil war. He also gave France the first successful constitution in its history (after twelve failures), which still serves well after half a century. These are substantial achievements. But it would be wrong to suppose that de Gaulle was ever a popular hero in the way Churchill was from 1940 on. A certain segment of the population, the same (roughly speaking) which had idolized Napoleon III, supported him fanatically, though not with warm affection. He had devoted followers, but no friends. On the other hand he was much hated, in all parts of the political spectrum. He was always a divisive, never a consensual, figure. This was the consequence in part of objective facts springing from the deep wells of French history, but at least as much to his adamantine character and the free rein he gave to it.

The first thing that struck you about de Gaulle was his appearance. It was extraordinary. He was over six foot six, enormously tall for a Frenchman. Tall men interested him. He was delighted with Lord Reith (six foot eight), the head of the BBC, whose incongruous height made him a savage loner. At Kennedy's funeral in Washington, de Gaulle demanded to be introduced to the lofty Kenneth Galbraith—"*Presentez-moi à ce grand homme-là!*" His height cut him off from the rest of mankind, or so he felt. He made a virtue of his singularity, turned it into a philosophy of life. As he put it: "Solitude was my temptation. It became my friend. What else could satisfy anyone who has been face to face with history?" It is important to note that de Gaulle's height did not give him grandeur. He was not far from being grotesque. General Spears recorded his impression on first meeting de Gaulle in the crisis of 1940:

> A strange-looking man, enormously tall: sitting at the table he dominated everyone else by his height, as he had done when walking into the room. No chin, a long, drooping elephantine nose over a closely cut moustache, a shadow over a small mouth whose thick lips tended to protrude as if in a pout before speaking, a high, receding forehead and pointed head surrounded by sparse black hair lying flat and neatly parted. His heavily hooded eyes were very shrewd. When about to speak he oscillated his head like a pendulum, while searching for words. I at once remembered the nickname of *Le Connétable,* which Pétain said had been given him at Saint-Cyr. It was easy to imagine that head on a ruff, that secret face at Catherine de'Medici's council chamber.

According to Malcolm Muggeridge, who had various opportunities to be close to de Gaulle during the war, he suffered from chronic halitosis, and knew it. This strengthened his habitual tendency to push people away before they had a chance to draw back in distaste. He was, literally, a repulsive person. He had acquaintances, some of whom he met regularly in specified restaurants, for purposes of discussion rather than camaraderie. But he had no *camarades*. To get close to somebody was a possible source of weakness. He demanded intense loyalty from his followers, and usually got it, but gave none in return. There was a revealing exchange with one of the most devoted of them, Jacques Soustelle, about Algeria, after de Gaulle returned to power in 1958. Soustelle said he was unhappy about de Gaulle's move away from the Algerian *colons,* adding, "All my friends are unhappy." De Gaulle replied: "*Alors, Soustelle, changez vos amis!*" After de Gaulle's death, I asked Soustelle if this anecdote was true. He replied: "Yes. He did say that. He was a terrible man."

This terrible man, capable of evoking heroism of the grandest kind but also of inspiring the sublimity of dread, was an intellectual. By that I mean he thought ideas more important than people. More particularly he thought the idea of France infinitely more important than individual Frenchmen, whose role in life was simply to work, fight, suffer and if need be die for France. Though determined from an early age to be a soldier, and devoting his life to it wholeheartedly, he was never successful at soldiering. This was possibly bad luck: he spent much of the First World War in a prison camp (a notoriously destructive experience, which aggravated his sense of solitude). In his late thirties he was still a captain. His only *patron* was Marshal Philippe Pétain, who befriended him on a number of occasions. He got him to give a course of three lectures on leadership at the Military Staff College. Everyone present was senior to de Gaulle, and the lectures, an astonishing display of historical knowledge, delivered without a note, were heard in stony silence, Pétain alone expressing his approval. They discussed, among other topics, the nature of the hero in war, and the need, on occasion, for the hero to disobey foolish or cowardly orders from above. He published them later in a slim volume called *Le Fil de l'epée* (*The Edge of the Sword*), and they are an uncanny adumbration of his later role.

De Gaulle put his ideas in print on other occasions, in specialist journals, and in a book, *Vers l'Armée de metier* (*Toward a Professional Army*), which had much to say about armored warfare. He had a brief and successful experience of commanding a scratch armored brigade after the collapse of France had already begun in 1940, but otherwise his soldiering was essentially intellectual—the attempt to propagate novel military ideas. De Gaulle had no more affection for writers and intellectuals than he had for any other group in society. But he recognized the important role they had always played in French history, and genuflected to it. He had no reason to like Jean-Paul Sartre (who hated him), but, as president, he honored France's best-known philosopher and always addressed him as *maître*. He made André Malraux his minister of culture, seated him on his right hand at cabinet meetings, and usually asked him to open the discussion on any really important topic. These were not mere gestures. On one point de Gaulle agreed with Churchill: in the end, words last longer than anything else. He would have agreed with Dr. Johnson too, in saying, "The chief glory of every people arises from its authors." This point came out markedly in a press conference in Paris which I attended sometime before he returned to power but when the question of the European Union was already *sur le tapis*. But before describing it, it is necessary to record that de Gaulle's body language was quintessentially French. If there is one gesture the French male is particularly attached to, and brilliant at, it is the shrug of the shoulders (French women do not do it). De Gaulle, with his great height and peculiar pear-shaped body, turned the shrugged shoulder into a work of art. At this press conference, de Gaulle said he was not inspired by the coal-steel "community" as the foundation of the future European Union. He continued: "For me, the materialism of Brussels is uninteresting. *Pour moi, l'Europe c'est l'Europe de Dante, de Goethe, et de Chateaubriand*." I interjected: "*Et de Shakespeare, mon général?*" He darted at me a look of intense hatred, then paused, reflected, and gave a classic demonstration of his monumental shoulder shrug, adding: "*Oui, Shakespeare aussi*." It is worth pointing out that this little exchange showed de Gaulle, with characteristic percipience, putting his finger on the real weakness of the European Union in the half century since—its degrading materialism and its total failure to build itself on the civilization which is Europe's essence.

Being an intellectual, however, and thinking people less important than ideas, de Gaulle was not a warm and friendly man, or a generous or forgiving or magnanimous man. He was quite unlike Churchill in these respects. He was an egoist on a monumental, indeed a superhuman scale, someone who raised selfishness or self-regard into a principle of life, and of government and of metaphysics, and who, by identifying himself with La France, also embodied the natural selfishness which is the salient characteristic of France and of French people generally. There cannot have been on this planet, in the last two centuries, a human being more difficult to deal with or who came so close to the abstract quality of intransigence. De Gaulle made himself particularly difficult to those in Britain in the summer of 1940 who befriended and helped him to set up the Free French movement, notably Churchill. Indeed, if there was one characteristic even more prominent in de Gaulle than his egotism and selfishness, it was his ingratitude. He saw gratitude as a weakness in general and, to a man in his position, as a fatal weakness. If a grateful impulse ever sprang up in his stony bosom, he stamped it out instantly. Ingratitude to his personal benefactors, and to France's allies and friends, was absolutely central to his weltanschauung and to his geopolitics.

There was also, in de Gaulle, a streak of ruthlessness quite alien to the Anglo-Saxon world of politics, and which belonged more to the Continental autocracies of the ancien régime than to the constitutional democracies of the West. He did not say "*L'État c'est moi,*" like Louis XIV. Certainly he thought he was the French state, from 1940 till the liberation of 1944. But he went further. He thought *La France, c'est moi,* too. Thinking this, he felt himself constitutionally empowered, and morally entitled, to act exactly as he chose provided he could convince himself he was acting in France's interests. Hard as he was on his allies, the people he was hardest of all on were the French. They were, from time to time, the victims of the blackest side of his nature. He could be, and occasionally was, cruel, vicious and terrible. In keeping absolute control of the Free French movement, first in London, then in North Africa, and in frustrating attempts to infiltrate it by its various French enemies, especially the Vichy regime, he did not hesitate to allow his agents to employ murder and torture. His ruthlessness was also evident in his reestablished regime from 1958 onward. Not only did he turn on the people in Algeria, army officers and civilians, who had helped to restore him to power (many of whom were killed, exiled or suffered long terms of imprisonment), he also, as president, used the full resources of the law to keep his critics under control. Under legislation passed in 1881, it is a crime punishable by fine and/or imprisonment to insult the French head of state. Under the Third Republic, over sixty-nine years there were only six persecution convictions under the act. In the fourteen years of the Fourth Republic there were only three. But in the decade of de Gaulle's rule, 1958–1968, there were 350, chiefly writers, cartoonists and publishers but also ordinary citizens who booed when de Gaulle's image appeared in movies, or shouted "Down with de Gaulle" as his car passed in the street. There is no question that de Gaulle, though never personally involved in these many prosecutions—three a month—both authorized and approved of this legal strategy.

Yet de Gaulle's rule in the Fifth Republic he created, though authoritative, was never authoritarian. The student revolt of May 1968, though it appeared to shake his presidency, culminated in a general election which returned an overwhelming Gaullist majority, the biggest vote of confidence ever recorded in a French government. When de Gaulle lost a referendum, he instantly resigned from office for good, though there was no compulsion on him to do so. His pride, his amour propre, was always more powerful than his love of rule. Like Churchill he wrote his memoirs. He finished his wartime series but only half-completed his account of the years after 1958. The prose is magisterial in places, like Churchill's, but the interest is less compelling. The egoist often elbows out the historian. He was less experienced than Churchill at this mode of self-presentation and had not been able to establish a corner in the documents. Yet his views were always interesting, not least about the future. André Malraux recorded his last conversation with de Gaulle in which the general spoke of Europe:

> I have tried to set France upright against the end of the world. Have I failed? It will be for others to say. We are certainly present at the end of Europe. Why should parliamentary democracy (involving as it does here in France the distribution of tobacco shops!) which is on its last legs everywhere, create Europe? Good luck to this federation without a federator! Why should a type of democracy which nearly destroyed us, and isn't capable of ensuring the development even of Belgium, be sacrosanct when we have to overcome the enormous difficulties which face the creation of Europe? You know as well as I do that Europe will be a compact between the States, or nothing. Therefore nothing. We are the last Europeans in the Europe that was Christendom. It was a lacerated Europe, but at least it did exist. The Europe whose nations hated each other had more reality than the Europe of today. Don't ask whether France will make Europe—we have to grasp that France is threatened with extinction through the death of Europe. Nothing is final, to be sure. Supposing France became France again? I have learned the hard way that putting France together again from the broken pieces has to be done over and over. But perhaps this time . . . I have done what I could. If we must watch Europe die, let us watch. It doesn't happen every day!

Both Churchill and de Gaulle were old-style national heroes, with the limitations of such. But both had a transcendental gift of reflecting upon the process of history, which is perennially interesting.

12

HEROISM BEHIND THE GREASEPAINT: MAE WEST AND MARILYN MONROE

In their long struggle to achieve equality of opportunity with men, women have had to make use of all the physical advantages which their sex gives them, in youth anyway. Dr. Johnson thought the balance fair: "It is right that the law should give women so little power, since nature has given them so much." But nature's power does not endure. Though women have now for centuries lived longer (on average) than men, their charms wane more swiftly. History is punctuated by the sad tales of beautiful women, once all conquering, who ended as destitute. Emma Hamilton was such a case. So was her contemporary Mrs. Jordan, the Regency comedy actress who enchanted Charles Lamb, and later, as the mistress of the future William IV, bore him ten children, only to be cast off into a miserable old age of want and loneliness.

Such examples are so common that they inspire pity rather than a sense of tragedy when we come across them. So we welcome the occasional lady who makes a shrewd assessment of her charms from the start, capitalizes on her strengths, keeps her weaknesses in check, and ends a long, rich life with money in the bank, men on tap and always a few jokes to the good. There are not many such—very few indeed—and therefore they deserve, and get, heroic status; certainly among women, and among the more discerning men too. The doyenne of this select group was undoubtedly Mae West. And her life and career make a neat diptych with the outstanding example of the prevailing frailty, Marilyn Monroe—though she, too, by an ironic twist of fate, occupies a heroic niche in history.

Mae West was born on August 17, 1893, in Brooklyn, then a suburban town, not incorporated as a New York City borough until five years later. She was registered as Mary Jane West. Her ancestry was English, from Long Crendon, in leafy Buckinghamshire (her mother, Tilly, had German blood). Her father was a prizefighter, "Battling Jack" West, also a private detective, taxi-cab owner and many other things. Her mother had wanted to go on the stage but had been forbidden to do so by strict parents. Now she lived vicariously through May or Mae, as she became, following the 1890s song:

Her Christian name was Mary
But she took the "R" away
She wanted to be a fairy
With the beautiful name of May.

This was the dodge adopted by West's contemporary "May of Teck," later Queen Mary, stiff as a poker, wife of King George V of England. Mary was formal, pious and good, May was intimate, sexy, nice. Gradually, West made it sexier, as a teenager, by spelling it "Mae." Her debut, aged seven, was as Baby May—Song and Dance, at a Brooklyn Sunday theater. She was on the stage nearly eighty years, with a few gaps. Beginning with Hal Clarendon's Stock Company, she trained at Ned Wayburns's Institute of Dancing and College of Vaudeville, "the Place Where Chorus Girls Are Taught to Dance and Sing from the Raw Material, and Made Ready for the Footlights." As a teenager she was Little Lord Fauntleroy (a photo exists), in Shakespeare and French farce, a murder victim in

Dr. Jekyll and Mr. Hyde, and in thrillers and melodramas. She was on the same bill in 1907 as W. C. Fields, then described as a "Humorous Juggler," and almost certainly worked with "The Astair Children," Fred and Adèle, and "The Marx Brothers," Julius and Milton (later renamed Groucho and Gummo). By the time she was twenty, she had toured in burlesque, danced in a Broadway production of the *Folies Bergère,* acted in *Vera Violetta* and and *A Winsome Widow,* also on Broadway, and made a solo spot for herself in vaudeville. She had also married, aged eighteen, and kicked out, a no-good called Frank Wallace, who made periodic reappearances in her later career, demanding handouts.

Wallace clearly helped to shape West's views on men. She set them down in her "Ten Commandments on Men and Other Things" (actually there are fifteen), discovered under the heading "Things I'll Never Do" by her biographer Simon Louvish (whose superb work is the prime source for this essay) in an undated paper in the Mae West archive at the Academy of Motion Picture Arts and Sciences in Los Angeles:

Things I'll Never Do

1. Take another woman's man. Not intentionally, that is. Even though all's fair in love and war, and it ain't no sin.
2. To be anything but myself at all times, publicly and privately, except on the stage or screen, for that's where acting begins.
3. I'll never cook, bake, sew, wash dishes, peel potatoes, eat onions or bite my nails.
4. Wear white cotton stockings or join a nudist colony.
5. I will never like opera, Number Thirteen, yodelling, cold spaghetti, rats, snails, men who shave their necks, or overripe bananas.
6. Care for people who whistle in dressing rooms or bounce cheques.

7. Play mother parts, sad parts, dumb parts, or a virtuous wife, betrayed or otherwise. I pity weak women, good or bad, but I can't like them. A woman should be strong either in her goodness or her badness.

8. Go nuts about classical music, sandwiches, cigar smoke, places that smell like hospitals, and black nail-polish.

9. Get excited over night clubs, contract bridge, fan dancing, bobby-sox, the Stock Market, badminton or bust-developers.

10. Be thrilled to death by orchids, anonymous love-letters, souvenir postcard folders, earthquakes, slave bracelets, or beds with hard mattresses.

11. Be bothered by Scotch money-lenders or boys who lisp.

12. Believe the worst about anybody without complete proof nor will I believe it's useless to struggle against fate—the phony!

13. Walk when I can sit, or sit when I can recline. I believe in saving my energy—for important things.

14. Write a story that is unsophisticated, because I believe that innocence is as innocence doesn't.

15. Marry a man who is too handsome, a man who drinks to excess or doesn't carry his liquor like a gentleman, a man who is easy to get or easily led into temptation—unless I do the leading.

Although this set out to be West's manifesto on Men, and she remembered what it was supposed to be before she wrote her fifteenth point, this is really a glimpse into her inner life, which reveals how comparatively unimportant was the role men (as opposed to Men) played in it.

We do not know what Mae West really looked like. We do not know her height for sure: once she became sassy (her word) she became skillful in having herself photographed in such a way as to make it difficult to judge her height exactly. She cannot have been much over five feet since she always wore heels as high as possible, provided they were reasonably safe. She gave her measurements as "38-inch bust, 38-inch hips and 28-inch waist" and said this had never varied much. The notion that she had very large breasts—which gave rise to the nickname a Second World War air-crew gave to their inflatable life jackets—is plainly false. She was not a natural blonde and always dyed her hair. Its natural color was light brown or dark blond, not brunette, as her enemies claimed. She had very good teeth, and in 1949 when she was fifty-six, said to an interviewer, Earl Wilson of the *Omaha World-Herald,* "I've got all my own teeth," and invited him to look into her mouth and down her throat, "an offer which I was too gallant to refuse," he wrote, reporting the fact under the headline: "MAE STILL HAS ALL HER TEETH." Emboldened, he asked her age. She replied: "I am over twenty-one."

From the moment when Mae West appeared in solo spots in vaudeville she specialized in suggestive jokes and dressed to exploit her sexuality. During the 1920s she became the personification of sex to theater audiences. She associated, for publicity purposes, with prizefighters, wrestlers and athletes, musclemen of all kinds, and sometimes with those reputed to be gangsters. Her ostensible choice in men did not change, and in her seventies and even eighties she still liked to be photographed with boxers, often admiring their physiques. But was it all a stunt? There is no evidence at all that West had a particular interest in sex or wide sexual experience or knowledge, or salacious tastes. In some respects she was prudish. She never had herself photographed in the nude or topless. She never used four-letter words or crude, explicit expressions. She never kissed onstage or on film. She devoted much of her life to derisory innuendos and double entendres of great variety and ingenuity, putting them across with remarkable skill. But she never told a dirty story as such, onstage or in private. Moreover, she went out very little. In the classic age of the speakeasy, a place she made much use of in her stage material, she was hardly ever reported to be in one or photographed there. She figured rarely in the gossip columns except for obvious publicity purposes. She did not frequent nightclubs or premieres, except her own. The reason seems to be that when not actually onstage, she was working. West was a writer. In vaudeville, where she grew up professionally, she was competing not only with trained seals and other performing animals, escapologists like Houdini, trick cyclists and giant strongmen, but with her own generation of inspired comics, such as Charlie Chaplin, the Marx Brothers, W. C. Fields, Buster Keaton and Stan Laurel. The competition was keen, and the need for fresh, first-class speaking material and funny scenarios was constant. Most of the comics had specialties, and if they could write their own jokes, did so. Chaplin always did. So did Laurel, both when he was on his own and teamed with Oliver Hardy. Groucho, Harpo and Chico Marx wrote their material until they could afford to hire the best gagmen in the business. West soon discovered she could write better jokes and dialogue than what was provided for her, and she gradually took over her show entirely.

She was writing as early as 1912, when her vaudeville career really got going. In 1921 she registered a playlet, *The Ruby King,* as her copyright at the Library of Congress. The next year she registered a play, *The Hussy*. There followed two sex plays, *Sex* and *The Drag,* described as "a Homosexual Play in Three Acts." Both were registered as by "Jane Mest," a pseudonym for Mae West. *Sex,* in its first production, opened at Daly's, April 26, 1926, and ran for 375 performances. *The Drag* had only two previews, in Connecticut. *The Wicked Act,* registered by Mae West under her own name, had nineteen performances at Daly's in 1927. She followed this with *Diamond Lil,* her greatest script, with an initial run of 176 performances at the Royale Theater in 1928. Other plays by Mae West included *The Pleasure Man* (1928), *The Constant Sinner* (1931) and *Frisco Kate* (1930–1931), which was never produced. During this period West wrote two novels, *Babe Gordon*, published in 1930, and *Diamond Lil,* 1932. A third novel, *The Pleasure Man,* was not published till 1975. Further plays included *Loose Women* (1933), *Clean Beds* (1936), *Havana Cruise* (1939), *It Takes Love* (no date), *Catherine Was Great* (1944), *Come on Up* (1945) and her new versions of *Diamond Lil* for revivals, from 1947 into the 1960s.

Many of these scripts incorporated work by other writers or by West under pseudonyms, especially "Jane Mest." West's movie career, which began with *Night After Night* in 1932, has a more complicated scriptology. In this first movie, West took credit only for "additional dialogue." *She Done Him Wrong* (1933) was adapted from West's *Diamond Lil. I'm No Angel* (1933) was mainly by West, but three other writers were credited. *Belle of the Nineties* (1934) and *Goin' to Town* (1935) were essentially by West. *Klondike Annie* (1936) was a movie of her unproduced play *Frisco Kate. Go West, Young Man* (1936) and *Every Day's a Holiday* (1938) were by West. *The Heat Is On* (1944) appears to be her only movie in which she had no hand in the writing, though one suspects she actually supplied material during production. The most interesting case is *My Little Chickadee* (1940), by far her best movie—which she wrote jointly with W. C. Fields, each crafting their own lines.

I have described West's writing in detail for two reasons. First, it enables us to picture the showbiz world in which she matured and flourished. It had more in common with the London stage in Shakespeare's day, or perhaps the early Victorian provincial theater as described by Dickens in *Nicholas Nickleby* than the financially sophisticated entertainment industry of the early twenty-first century. West had no authorial amour propre or artistic pretensions. Everything was done ad hoc, much of it at the last minute. The only criteria were applause and box office. Second, and more important, West's role as author and gag writer enabled her to keep complete control of her showbiz personality and to present it entirely as she wanted. Between the wars she created "Mae West," an overwhelming showbiz image, one of the most famous and enduring in the world, by devising her situations and writing her lines. No other female star of her time, probably of any time, was so completely her own creation. Just as Garrick was Garrick, so Mae West was all her own work.

The results justified the effort that went into the creation. When West shifted from stage to screen early in the 1930s, taking full advantage of the new talkie process, she became not just a New York star but world famous. And she was paid accordingly. In 1934 she became the highest-paid entertainer in America, with an income of $399,166.00. The next highest was W. C. Fields, with $155,083.33. The only other woman performer in West's league was Marlene Dietrich, with an estimated $145,000. In 1935 West earned more than any other female in the world, with $480,833. Only one man exceeded her take-home pay, William Randolph Hearst, with an estimated $500,000. The next three in the list of top earners were Alfred P. Sloane, president of General Motors, with $374,505, Marlene Dietrich with $368,000, and Bing Crosby with $318,000.

West made her fortune by working outside the Hollywood studio system, which kept actors in gilded slavery under long, exclusive contracts giving them no share in profits. By making herself her own boss, she won both freedom and affluence. She invested her earnings in sound stocks and bonds at a time in the 1930s when you could get in at the bottom of the market. But above all she bought high-quality diamonds, which served as her chief stage prop. By the end of the 1930s, West was probably the most celebrated owner of precious stones outside India. She wore them too, as often as possible, leading to her best spontaneous one-liner when an usherette at the Hollywood Roxy exclaimed at a premiere in December 1936, "Goodness, what diamonds!"—"Goodness has nothing to do with it."

The only effective restraint upon West's artistic freedom, both during her twenties Broadway career and her thirties Hollywood epoch, was censorship. In New York West was targeted by the city authorities and by Aimee Semple McPherson, then leading a moral crusade there. In April 1927 West was convicted of corrupting public morals for writing, putting on and acting in *Sex,* and sentenced to ten days in jail plus a fine of $500. She served her sentence, actually eight days, since she got two days off for good behavior, on Welfare Island, a former insane asylum. The New York *Daily News* carried a report on her experiences every day, and she herself wrote a full account in *Liberty* magazine (August 1927). She wove her prison term skillfully into her public image, adding an extra dimension by arguing that her plays, by informing ignorant women about the dangers of sex, kept down the rate of illegitimacy and venereal disease. She gave her public-interest dimension a proto-feminist twist by emphasizing that all the lawyers in the case were men, that the jury was all male and that no women witnesses were called. She said: "This was a case of Men *versus* One Woman." She took the same line during her movie career in the 1930s when she ran into constant trouble with the Hays Production Code.

West staged an elaborate comeback in her seventies, based on a film version of Gore Vidal's story *Myra Breckinridge* (1970), which has since settled down to a long career as a cult movie particularly favored by homosexuals. There was also a 1976 movie, *Sextette,* based upon a play she wrote in the 1950s. In the same decade there were nightclub acts and appearances in Las Vegas, for which West wrote the scripts, and radio and TV appearances, of which Simon Louvish gives a list in an appendix to his biography, though he says it is certainly incomplete. Her last live appearance, at the San Francisco premiere of *Sextette,* was in 1978, when she was eighty-five. She died two years later, on November 12, 1980, and was buried in the family vault in Brooklyn.

Over the three-quarters of a century during which West was a professional comedienne, she built up one of the most consistent and indestructible personas in the history of show business. It was all her work and remained throughout under her control. It was punctuated by songs and wisecracks that reinforced the consistency. The songs, partly and sometimes wholly written by her, tell their own story: "They Call Me Sister Honkey Tonk," "Willie of the Valley," "I Wonder Where My Easy Rider's Gone?," "He Was Her

Man, But He Came to See Me Sometimes," "A Man Who Takes His Time," "He's a Bad Man, But He Treats Me Good," "What Do You Have to Do to Get It" and many more. West also gave unusual, and characteristic, renderings of classic songs, such as "Silver Threads Amidst the Gold."

She worked hard on her gags throughout her life, and her enormous gag book was, next to her diamonds, her most precious possession. "Come up and see me sometime," she never uttered until it was world famous. Like other classic movie lines—"You dirty rat," "Play it again, Sam," "I want to be alone"—it was invented, or, rather, polished, by the public. West was in the great American tradition of one-liners, which goes back to Benjamin Franklin. "A hard man is good to find." "It ain't the men in my life; it's the life in my men." "A man in the house is worth two in the street." "What good is alimony on a cold night?" "A girl climbs the ladder of success rung by rung." "Beulah, peel me a grape." "When I'm good I'm very good, but when I'm bad I'm better."

West began her gag collection, now in her archive, in the early 1920s (perhaps based on notebooks going back to 1907), and it eventually numbered two thousand pages, contained in over forty folders. There were twenty thousand jokes altogether. Much is typed, but five hundred pages are in her own handwriting, leaning backward precariously. She used stock publications written for the use of stand-up comedians, such as *Madison's Budget, McNally's Bulletin* and *Digest of Humour*, some of them going back well into the nineteenth century. But the vast majority were originals or versions of her own. An enormous amount of work went into this project, and there is something touching about the way this self-disciplined and industrious woman spent her life looking for the perfect joke, while making do with imperfect ones. If she set out to prove that a woman can outdo the men in the grand and grim task of show business, she certainly succeeded.

West's appreciation of the fact that women see sex as intrinsically linked to jokes, whereas men tend to take it with intense seriousness, helps to explain why she was so popular with women. More than two-thirds of her fan mail was from women. She was a heroine to the younger generation of actresses who followed her, chiefly because of her success in controlling her own career, and her ability to orbit fame for so long outside the star system. This admiration built up long before the opening of her archives revealed the full extent of her writing activities and her robust creative vitality. One movie actress who admired her (she told me so) was Shelley Winters. "I came from Brooklyn too. I was taller than Mae and didn't have to wear two-inch heels. But I had no natural assets. At Woolworth's, my first job, I was put in hardware because I wasn't pretty enough for candy. My figure grew voluptuous by eating. In my early movies I was a victim, always. I was drowned by Montgomery Clift, run over by Alan Ladd and James Mason, knifed by Robert Mitchum and strangled by Ronald Coleman." Her first movie was in 1943, and after the war, when she was getting known, she had an encounter with Dylan Thomas, which has gotten into the literary anecdote books in various versions. Winters's own was: "He told me he had come to Los Angeles to touch the titties of a starlet, as he put it. I said, OK, but only one finger, and dip it in champagne first."

As a starlet, Winters shared an apartment for a time with Marilyn Monroe, who was six years her junior. They had two "good" outfits which they shared: a swimsuit for photo shoots and a mink for dates. She taught Monroe how to "look pretty" by tilting her head back, lowering her eyes and parting her lips. They listed the men they would like to sleep with or marry. Monroe's list included Albert Einstein, Arthur Miller and other intellectuals. "I said, stick to movie actors, it's safer." She followed her own advice, and bedded Marlon Brando, William Holden, Burt Lancaster, Errol Flynn and many others, made over 150 movies, died comfortably rich, in her own bed, at eighty-five, having won two Oscars and survived three marriages.

Winters provides a bridge between the two contrasting lives of

West and Monroe. She saw both as heroines, West the boss-heroine, Monroe the victim-heroine: "I admired West but I wanted to be Marilyn—she had all the assets." West agreed. She said of Monroe: "She was the only girl who ever came close to me in the sex department. All the others had were big boobs." It is probably true that, by the standards of commercial display (an economist's name for showbiz), Monroe had the finest body ever recorded, from top to toe. But she was lucky to have been born in the 1920s—at least she had a life, or thirty-six years of it. If she had been conceived in the twenty-first century, she would, without much doubt, have been aborted, and that remarkable body would never have come into existence.

Her mother, maiden name Monroe, generally known as Gladys Baker, was a pathetic creature who hovered most of her life on the brink of sanity and spent more than half of it in institutions, in one of which she died. Her mother had also died in an asylum, in a straitjacket. The males of the family were alcoholics, when they were on the scene at all. Monroe was registered at birth as Norma Jean Mortenson, though Mr. Mortenson was almost certainly not her father. Her mother, when not institutionalized, worked in a film lab and her daughter always recognized her "because she smelled of glue." The father may have been a man in the same lab, name now unknown. Monroe never met him, or knew who he was for sure. Her mother's role in her life was fleeting and blurred. The nearest Monroe got to a parent was her mother's friend Grace McKee, who also worked in the film lab and was known as "Aunt Grace." But she could not, as a rule, look after the child, who was farmed out to a series of semiprofessional foster parents, eleven in all. On September 13, 1935, McKee, having run out of foster homes (in at least two of which the pretty child was sexually abused), deposited Monroe, aged nine, at the Los Angeles Orphans Home on North El Centro Avenue. Monroe had to be dragged inside, protesting she was not an orphan. She slept on a small metal bed in a big dormitory with twenty-six other girls. In the dark she could see through the uncurtained window at the bottom of the room the flood-lit and illuminated water tower of the RKO Studios on Gower Street.

After that experience, Monroe, thanks to much pleading, was taken in by "Aunt Grace," who gave her her first lipstick and rouge, and taught her to use it, aged ten. Aunt Grace had just married her fourth husband. He drank, and tried to rape the child. So it was back to foster parents again. When abused, as she repeatedly was, she was blamed for being "provocative." At fifteen, a young man called James Dougherty, then twenty-one, showed an interest in her and offered to marry her to take her off Aunt Grace's hands. She was told she could either marry him or go back to the orphanage. So on June 19, 1942, three weeks after her sixteenth birthday, Norma Jean Baker was married. He joined the U.S. armed forces and left, and she lived with his parents. Monroe got jobs in war factories, as a chute packer, paint sprayer, and so on, and in one of them was spotted by an army magazine photographer, who advised her to take up modeling. He said something interesting: "You were born to be married to the camera." There followed the Blue Book Modeling Agency, a screen test at 20th Century Fox, her first brush with the fierce producer Darryl Zanuck, who cast an angry shadow over her life, and work as a film actress.

It would be hard to imagine a more depressing and hostile background than the one which encompassed Monroe, composed as it was (I have left out many of the horrible details) of severe mental illness, syphilis, alcoholism, divorce, illegitimacy and desertion. She was also left-handed (as were Betty Grable, Judy Garland and Olivia de Havilland) at a time when to be a sinistral at school, as I know from experience, was much more of a burden than it is now. On the other hand, she had beauty and health. Moreover, she had a species of magic of the rarest kind, an ability to interlace herself with the process of photography, whether still or motional, to a degree no other actress has ever possessed. The camera is an extraordinary machine. It is like the printing press. We think of it as a mere human contrivance to make the telling of truth easier, simpler and cheaper. But it is not by nature a truth teller. It is a transformer, a self-generating art form on its own, often escaping from human control and pursuing its own aims; or taking possession of human beings and using them. The camera is more dangerous and deceptive than the press, because less well understood. We all know, and have known since the 1450s, that the press can not only reproduce lies but enlarge them. But the camera has posed since its inception in the 1830s as the antipode to the pen and brush because of its mechanical inability to reproduce anything except what it sees. Hence the phrase, dating from the mid-nineteenth century, "The camera cannot lie." No greater untruth ever came into popular usage. The camera lies all the time, through an infinite spectrum of mendacity. It is a mechanical monarch, with its favorites and its enemies list that stretches to the crack of doom. Its favors are distributed by the system of rules but in the most capricious and arbitrary manner, which is usually as incomprehensible to its recipients as it is to those who operate the mechanism. It belongs to the world not so much of pleasure but of metaphysics.

Marilyn Monroe was the spoiled princess of celluloid, but she never recognized or understood the birthright. On the contrary: her life was a constant struggle not to be possessed and raped by the camera, punctuated by rare moments of submission when she engaged in lovemaking with the merciless machine. She never understood that it loved her, let alone why; and throughout her professional life she felt that it was not just pursuing but persecuting her. She was shy to the point of mania. She appears to have been one of those rarities who suffers from both claustrophobia and agoraphobia, sometimes at the same time. For her, appearing in the morning on time to begin filming required an effort of courage akin to that demanded of a soldier in the First World War who had gone "over the top" too many times, whose reserves of courage had been exhausted, and who was now asked to do it again with nothing but fear in his belly. Each time, it required an act of valor physical, mental, even spiritual, and each time it became more difficult.

In short, Monroe's fifteen-year career as a film actress was the saga of a fight to the death over her magical body. On the one hand there was the camera, which lusted after it and deified it. On the other was the fragile, terrified woman who saw it as a devouring monster, and tried to escape it. If she had understood the nature of the struggle, all might have been well. But all she grasped was her fear. There is no doubt about the camera's lust for her. It propelled her upward throughout her career. She started as an extra. So did Dietrich, Jean Harlow, Merle Oberon and Sophia Loren. These actresses had to work their way upward. In Monroe's work as an extra she tended, involuntarily and unconsciously, to upstage those with speaking parts. Given a small speaking part, as in *Gentlemen Prefer Blondes,* she upstaged the principals. She was, unknown to herself, a movie witch. There was something unfathomable about her magic. She could look horrible: hair greasy and lank, eyes dull, skin unhealthy, figure recalcitrant, movements uncoordinated. Then the camera poured its love on, and transformed, her. When she first appeared before Laurence Olivier to begin work on what became *The Prince and the Showgirl,* he thought her so awful that he mentally decided to abandon the project there and then. Then he saw the rushes of her test, and embraced the movie with enthusiasm. (A huge mistake, as it turned out.) Many of the most beautiful stills of Monroe were taken at a time when she was described as "hideous." The photographers themselves could not believe their eyes when they saw the prints. The famous nude camera shot of her earned $500; within ten years the company that owned the rights had netted $1,000,000, though nude shots of starlets were common enough. The famous skirt-blowing scene over a hot-air grating, filmed for *The Seven Year Itch* on September 15, 1954, on the spot and witnessed by over two thousand people, causing traffic jams and great hilarity and anger, became one of the most notorious episodes in the entire history of the movies. In fact it had been tried countless times before, with actresses of all kinds, and always ended up in bad taste if not downright awkward vulgarity. With Monroe it worked perfectly, becoming not only graceful but even moving.

Monroe's fear of the camera, on the other hand, probably sucked up and regurgitated all the horrors of her miserable childhood and adolescence. Anyone writing about her is tempted to indulge in amateur psychiatry, a temptation which will be resisted here. She had little effective education but enough to make her yearn for more. There was an aching hole in her which needed to be filled. Religion might have served her well. But she was God resistant. When she and Jane Russell were working together, Russell, who was a Christian fundamentalist (of a sort), suggested to her: "Why don't you come with me after work to a Bible reading?" The answer was: "I'm having a reading to myself, with Freud." Monroe brings to mind G. K. Chesterton's reputed saying: "If people cease to believe in God, they don't believe in nothing, they believe in *anything*." Monroe had a hunger for the cerebral mysteries, as her conversation with Shelley Winters suggests. She had no wish to get close to male movie actors, though she occasionally did so, faute de mieux. It is significant that the two greatest Hollywood predators of the age failed to make her. George Raft, so successful and insatiable that he once made love to seven chorus girls in a single night, was rebuffed. So was Zanuck, who never forgave Monroe for giving him the brush-off, and did his best to derail her career. He had himself quoted as denying hotly that he had made a pass at her: "I *hated* her! I would not have slept with her if she'd paid me." On the other hand, any absentminded professor was always in with a chance. Monroe grew up a genuine pseudointellectual looking for a tutor. It is a thousand pities she never met Wittgenstein.

What she did meet were Method actors and the people who manipulated them. That was disastrous. Though Monroe had terrible fears and anxieties, she was by nature a natural actress, up to a point. A sensitive and sensible director would have gotten her to perform adequately and that, combined with her camera genius, would have been enough. Unfortunately, the Method people persuaded her that, if she learned from them, she could become one of the greatest actresses who had ever lived. She believed this as an article of faith. Indeed, Method became her religion. This was complicated by her emotional life. Loving celebrities, as she loved intellectuals—as a Platonic ideal—she chose as her second husband the sports star Joe DiMaggio. The marriage was unsatisfactory, as any marriage with her was likely to be. But at least he did not do her positive harm. Her next marriage, to the left-wing intellectual playwright Arthur Miller, enormously complicated her professional problems, and pushed her further into the grip of the Method people. Living up to the genius of Miller, and living up to the impossibly high ideals of the Method—as she saw them—redoubled her terror at facing the cameras, and the agony of acting in front of them. She retrogressed. From being occasionally late in the morning, she became habitually late. From being thirty minutes late, she became an hour late; then two, three, four hours late. From forgetting the odd line, she forgot them all. Her dependence on medication and alcohol increased pari passu.

Whether Monroe would have been a more professional performer without her Method obsessions is a hypothetical question which cannot now be answered. The camera remained faithful to her to the end, and her appeal to the public continued to increase. But her rating within the industry fell. Producers began to dislike her because her antics increased costs. Directors found she added an extra dimension of difficulty to making a movie, and they resented her insistence that a Method coach be on the set and give her advice between takes. Other actors thought her selfish even by theatrical standards, and damaging to their interests.

One victim was Laurence Olivier, with whom Monroe filmed *The Prince and the Showgirl* in 1956. He was then regarded as the world's greatest living actor and was at the height of his fame as a director. Monroe might have learned a lot from him but her Method coach came first and Olivier was offended and paralyzed by this alien presence during filming. An intelligent, even cunning, theatrical operator, he made a fundamental mistake early on in the production, which wrecked his relationship with her. Just before a key take, he whispered: "Be sexy." This was otiose advice since, on form, Monroe was incapable of being anything else, but it was a vulgar insult to an actress aspiring to be a combination of Eleonora Duse and Ellen Terry. She flounced off to her dressing room in tears, and thereafter it was all downhill. At the wrap-up, she apologized in front of all the cast for being "so beastly. I hope you will all forgive me. It wasn't my fault. I've been very, very sick all through the picture—please—don't hold it against me." But they did of course. The movie was a flop and no one emerged with credit, except Monroe, who made some blissful epiphanies.

Even more disastrous for those involved was the 1960 movie *The Misfits,* which Monroe made with Clark Gable, king of the stars in Hollywood's golden age, and still capable of radiating a sex appeal on film akin to her own. Her marriage to Miller, who wrote the script, was breaking up, and she was at her worst. An on-the-set book, *The Making of the Misfits,* gives a blow-by-blow account of the horrors, which made the movie the most expensive black-and-white picture ever made at that date. Clark Gable complained bitterly at her unprofessionalism, though he admitted he liked her. There is a charming still of them hugging each other at wrap-up time, which says everything about the photogeneity of both. But Gable said to a friend a few moments later: "I'm glad the picture's finished. She damn nearly gave me a heart attack." He spoke too soon. The next day he had one on a massive scale, and ten days later he was dead.

Before this tragedy, Monroe had made her masterpiece, *Some Like It Hot,* under Billy Wilder, the only director who ever agreed to work with Monroe twice. The screenplay is superb and Wilder got the best out of perhaps the most brilliant cast ever assembled in Hollywood. But it is Monroe and her beauty which casts a magic glow over the picture, and has made it perhaps the most loved of all movies. Like that other subtle masterpiece, Jane Austen's *Emma,* it reveals new merits each time it is enjoyed, not least in Monroe's projection of herself. It was a case of all's well that ends well with this picture, and Wilder, to the end of his days, used to delight listeners with anecdotes of how he contrived to get Monroe to appear on the set and remember her lines. One, "Hello, it's Sugar," required forty-one takes. But one actor, Tony Curtis, remembered the filming with bitterness. He said that when he and Monroe played a scene together, Wilder was never content until he got a perfect take from Monroe, and went on and on, while her takes got better, and his own worse. The printed take was always her best and often his worst. He got his revenge by claiming they had had a brief affair during the filming: "And kissing her was like kissing Hitler."

Monroe's last period in the early 1960s ended with her being sacked from the movie *Something's Got to Give,* and her overdose. It was complicated by her involvement with the tawdry "Camelot" circus of President John F. Kennedy and his vulpine brother Robert, then the attorney general. She seems to have had affairs with both men, though not, happily, with the third brother, Edward, so did not share the pathetic fate of Mary Jo Kopechne. Her relationship with the Kennedys is memorable if only for her appearance to sing "Happy Birthday" at a celebration rally for the president, then at the height of his ephemeral celebrity. This was the last epiphany of the star the camera worshiped. But it is impossible to write adequately of this tragic heroine, raised from nothing to sublimity and then dashed into dust. Perhaps heroine is the wrong word. But then what is the right one?

13

THE HEROIC TRINITY WHO TAMED THE BEAR: REAGAN, THATCHER AND JOHN PAUL II

Three people won the Cold War, dismantled the Soviet empire and eliminated Communism as a malevolent world force: Pope John Paul II, Margaret Thatcher and Ronald Reagan. They worked in unofficial concert and we shall perhaps never know which of the three was the most important. John Paul effectively undermined the Evil Empire (Reagan's phrase) in its weakest link, Poland, where the process of disintegration began. Margaret Thatcher reinvigorated the capitalist system by starting a worldwide movement to reduce the public sector by a new term, "privatization," and by destroying militant trade unionism. Reagan gave back to the United States the self-confidence it had lost, and at the same time tested Soviet power to destruction. All were heroes, each in a different way.

Reagan interested me the most because he created an entirely new model of statesmanship: well suited to a late twentieth-century media democracy. And he was hugely entertaining to watch in action. He endeared himself to me the first time we met by getting flustered, glancing at the six-by-four cue card he always kept in his left-hand suit pocket, and saying: "Good to see you again,

Paul." The second time he shook hands with me in front of a battery of press photographers (I still have the picture) and whispered: "Don't look at me—look at the cameras." Good advice from an old pro. Reagan did not try to smile all the time, like many American politicians. He never smiled at nothing. His smile was an event with meaning, which preceded or followed a joke. Usually he was serious. Government, he seemed to say, was a serious business. So serious we're inclined to take it too seriously. Then would follow a joke, and a laugh. But even when emphasizing the seriousness of it all, Reagan never gave the impression of being nervous, or gloomy, or worried. He was at ease with himself. I have never come across a person, certainly not in public life, who was so thoroughly and fundamentally at ease with himself. By that I do not mean casual or flippant or devil-may-care: he was none of those things. But he was relaxed, unharassed, quietly confident in anything he had to do. And, being like that, you did not have to dig very deep to find happiness. He was a happy hero. He liked, and tried, to communicate this happiness, and normally succeeded. He made me think that happiness ought to be part of the equipment of a hero, even though it usually isn't.

The United States which Ronald Reagan took over early in 1981 was not at ease with itself. Indeed, it was deeply unhappy at a public level. The strong presidency of Richard Nixon had been destroyed, leaving a vacuum of power. Into that vacuum stepped, insofar as anyone or anything did, a divided and leaderless Congress, abrogating to itself by law or in fact duties which rightly belonged to the executive branch. President Jerry Ford did nothing to stop this unconstitutional larceny. He had never been elected and did not have a sense of rightful authority. He was not at ease with himself—far from it—but he was easygoing, diffident, amiable, anxious to avoid rows that might end in a challenge to his credentials. His successor, Jimmy Carter, was a natural one-term president at a meager time, who found it impossible to strike a national note. Both men ran a low-key presidency, stripping both the White House and its internal motions of any element of grandeur. Ford stopped the Marine band playing "Hail to the Chief" when the president arrived. Carter let it be known that he worked in the Oval Office in a sweater, and he encouraged his staff to "dress down" (the first time the phrase was used). All ceremony associated with the White House and presidential movements was cut to the minimum. Gradually the heart of American government acquired a slipshod air. Ford was a non-hero, Carter an anti-hero. "Jimmy," as he liked to be called, despised heroics, or said he did, and anyway, was incapable of them. His was a painfully unheroic presidency, culminating in a humiliating disaster to American arms in Iran.

From the start, Ronald Reagan reversed this process of American self-effacement. In his eight years as governor of California, he had raised the administrative profile of the state with the world's eighth-largest economy above the usual seedy city hall level. Now, entering a new role, he determined to play it to the full. He had the best of all precedents in stressing the formality and even the grandeur of the most important elective office in the world—the example of George Washington. He also had an able and enthusiastic assistant, in the shape of John F. W. Rogers, the young official in charge of White House protocol and ceremony. Rogers was an expert on everything to do with presidential history and all that was most seemly. He provided the costume and sets, as it were, for the Reagan presidency. Back came the solemn band music and especially "Hail to the Chief." Back came the Herald Trumpeters, from the U.S. Army, an institute created by General Eisenhower in his White House term. A special ceremonial fanfare was created for them entitled "A Salute to a New Beginning." Under Reagan's benign approval, Rogers rewrote protocol for all White House formal occasions, stressing ceremony, even redesigning the bunting used on presidential platforms. There was huge reviewing of troops by the president personally. All visitors, especially heads of state and government, were now suitably greeted. The internal dress code of the White House went back to "smart"—suits, ties, white shirts.

So Reagan began his rule by putting back the clock in a visible, audible way.

It should not be thought that Reagan's political heroism was entirely histrionic, though the theatrical element was of great importance in its success. But it could not have succeeded at all if Reagan had not been such a deeply serious man. He had certain core beliefs in which he passionately believed, from which he could not be budged, and which had a bearing on all he aspired to do. They were essentially moral beliefs, to do with justice, honesty, fair dealing, courage and what he would call "decency." In political terms they translated into standing up to the Soviets and matching them—if possible, outmatching them—in weapons; cutting taxes; freeing Americans from unnecessary burdens; and enlarging freedom whenever consistent with safety and justice. There was no shifting Reagan on these matters. He clung to his core views with extraordinary obstinacy. They were, by and large, right, and he could communicate them with extraordinary skill.

After nearly sixty years of writing history, and also of observing contemporary history makers in action, I am convinced that successful government depends less on intelligence and knowledge than on simplicity—that is, the ability to narrow aims to three or four important tasks which are possible, reasonable and communicable. Reagan had that formula, and the fact that he did enabled him to be a success, and a true hero, with few if any of the qualities which most constitutional experts would have rated indispensable.

Reagan was superficially, and also profoundly, ignorant. He did not seem to know how bills were put together or passed through Congress, or how the entire budget process took place. He had little education, and no desire to acquire much more in a general sense, at any rate through books. He was intellectually lazy, and he did not read one word of the carefully prepared briefing book on the eve of the world economic summit in 1983. When upbraided by James Baker he said calmly: "Well, Jim, *The Sound of Music* was on last night." The man who wrote the best book about his presidency, Lou Cannon, calculates that during his presidency he spent more time watching movies than doing anything else: he saw 356 at Camp David alone. Most of his history he derived from old Hollywood epics. Science fiction, or what General Colin Powell called his "little green men," was particularly real to him, one reason he opted for the Star Wars policy and defended it so stubbornly. It turned out to be one of his most successful initiatives, doing more than any other to break the will of the old Soviet regime to compete with America on advanced military technology. He had such insights, which were almost metaphysical. They were supplemented by nuggets he dug out of *Reader's Digest* and an inflammatory right-wing publication called *Human Events,* which aides tried to keep out of his hands.

He puzzled his staff. Sometimes he displayed extraordinary scraps of knowledge about obscure events. Sometimes he believed in fantasies, such as that the United States really had much larger hidden oil reserves than the whole of the Middle East. At other times he appeared incapable of speaking coherently about the simplest matters without reference to the cue cards in his left pocket. In some ways he was ill-equipped to run anything, let alone the mightiest nation on earth. He was nearly seventy when he got to the White House, and three months later an assassin's bullet just missed his heart. He was deaf and sometimes could not hear what his staff was telling him, even with the volume of his hearing aid switched right up. He was known in showbiz as "a quick study," and as a rule learned his lines well. As a B-movie actor, and a successful and reliable one, he had been a stickler for strict studio discipline, disliking people who were late on the set as "disruptive" and "unprofessional." He believed in learning lines and following the script, and obeying directions, so that in some ways his staff found him very compliant and easy to work with. But when tired, as he often was, especially after lunch, he got things wrong. He told a fund-raising meeting: "Now we're trying to get unemployment up, and I think we're going to succeed." He confused names and faces.

He thought his own secretary of commerce was a visiting mayor. He believed Denis Healey was the British ambassador. He addressed the Liberian president Samuel K. Doe as "Chairman Moe." But if he got the details wrong, he always got the mood right. It was rare for a visitor to the White House to go away unhappy, even if he got nothing that he wanted. Reagan was perhaps the most accomplished of all presidential performers, outdistancing that magician Franklin D. Roosevelt. He deliberately drew on his theatrical skills, as well as using them unconsciously. He said frankly: "There have been times in this office when I've wondered how you could do the job if you hadn't been an actor."

Reagan had two endearing characteristics which were also important to his success, and to his heroic status. First, he was genuinely modest. He had come from nowhere and had a perfectly proper pride in the things he had done, but he never pretended to be bigger than he was or know more than he did. He was not afraid to ask elementary questions. He asked for places to be pointed out to him on the map. "Where is Sri Lanka? Is that how you pronounce it?" "Who was this Atatürk?" "What was the Balfour Declaration?" He asked questions like "Why are the Blue Ridge Mountains blue?" and "Where did the English language come from?" "What was funny about *The Divine Comedy*? Why did they call it that?" Asking simple questions sometimes requires courage, but a man running a big, complex government needs to ask them.

Reagan did not pretend to know all about government. He never took refuge in its mysteries, though, God knows, they were often mysteries to him. He adopted an old trick of Andrew Jackson's, to distance himself from its processes. He referred to it always in the third person, thus ranking himself firmly with the "we" as opposed to the "them." Like Lincoln, he communicated with the people by a kind of cunning simplicity. Of course he did not have a powerful analytical mind like Lincoln's. He would never have been capable of the kind of sustained argument Lincoln produced in his great 1858 Douglas debates, or the sublime, deeply felt rhetoric of the

Gettysburg Address. For the grand occasion Reagan had to have his lines written for him. But just occasionally he had the Lincoln touch. "The big decisions are simple," he said. "That doesn't mean they're easy."

Where he resembled Lincoln was in his use of humor. Each knew humor was important in ruling, but each was sui generis in the way to employ it. Lincoln usually elaborated or invented his jokes, which might be subtle and complex—often thoughtful—and he used them as Christ used parables. Reagan used jokes as a substitute for logic and intelligence (and factual knowledge). It was part of his strategy for sidestepping arguments requiring exact information. He worked through metaphor and analogy, and humor. He did not possess, like Mae West, an immense baggage train of twenty thousand jokes, written by him or accumulated over many years. But he did have a stock of one-liners, about two thousand of them, which could be made to fit virtually any situation that confronted him as president.

He used jokes in the first place to put people at their ease. Laughter is the great reassurer and the great equalizer. A good example occurred when he was sitting in a helicopter next to a very nervous woman staff member, as it flew low over Washington. He said: "That's Arlington Cemetery. How many dead are there?" "I just don't know, Mr. President." "They all are." The tone of voice, the timing, was perfect, as always. Whether Reagan invented the joke, and this was the first time he told it, we cannot know. He certainly could invent jokes. When, immediately after the attempted assassination, and he was in considerable pain and was wheeled into surgery, he said to the anxious team of doctors: "I hope you're all Republicans." And when his wife was first allowed to be at his side: "Honey, I forgot to duck." (This was the use of an old 1930s catchphrase.)

No politician ever told stories (his or others) better. Lincoln's were shrewder and more original, but Reagan's were just as effective, perhaps more so because the tone and timing were more professional. He could get a laugh out of nothing. He destroyed Jimmy Carter with one quiet remark, after a characteristic Carter tirade: "There you go again." Most of Reagan's political one-liners centered on economics. He knew little about the subject and told jokes not so much to hide his ignorance as to share it with others. "Government does not solve problems. It subsidizes them." "Economists are people who see something works in practice and wonder if it would work in theory." "A recession is when the next guy loses his job. A depression is when you lose yours. Recovery is when Jimmy Carter loses his." Sometimes a Reagan one-liner enshrined a profound truth: "I'm not too worried about the deficit. It's big enough to take care of itself." I witnessed Reagan's first use of this joke, and the double take as those listening grasped its meaning.

One of the most dangerous things anyone can do, especially a politician, is to tell jokes against himself. For a clever, secure, witty person, who moves in sophisticated society, it is a tempting thing to do, but almost invariably damaging, since the admissions are taken out of their warmhearted context and used against you later. The great Lord Curzon was a notable victim of this endearing propensity. Reagan was the only politician I have ever known or heard about who not only told stories about himself, but used them deliberately and systematically to disarm criticism. His advanced age, his laziness, his stupidity, his ignorance and his general unfitness for office were all turned by him deftly into jokes. He thus took the detonator out of the most telling criticism before it could be fired. Indeed, he turned his weaknesses into virtues. Thus, on his age, he loved to explain how wage and price controls by government had never worked ever since they had first been tried under the Roman emperor Diocletian, adding: "I'm one of the few persons old enough to remember that." He emphasized his mental limitations by autographing old photos of himself when he was acting with Bonzo the Chimp, putting below his signature: "I'm the one with the watch." He turned self-deprecation into an art form by adding a dimension of gentle humor, turning defects into assets and persuading people he had the priceless gift of self-knowledge and self-criticism.

This was part of his way of bringing himself down, or the public up, to his level of importance, and so making a popular dialogue not only possible but easy and genial. Americans demand of their president both stature and a certain egalitarianism, both greatness and conviviality, no easy combination. They want to shake warmly the hand that, in national contexts, they want raised in admonition, warning and command. With his jokes, Reagan brought people to him while still keeping them at a respectful distance. No one—some said even his wife Nancy—got really close to him. His staff was warmly appreciated, but then, in the natural course of events in politics, discarded with grateful thanks, and without a pang, forgotten as swiftly and completely as last week's briefing.

A term applied to Reagan was "warmly ruthless." He was agreeable always and at times spectacularly charming. There is a telling memoir about him by his chief of protocol, Selwa "Lucky" Roosevelt, which shows the charm in action. For instance, when he reappointed her for his second term he sang to her the delightful song from *My Fair Lady*: "I've grown accustomed to your face." But she also provides a vivid account of the destruction of General "Al" Haig, when that important and self-important official had outlived his usefulness. Reagan's favorite remark to his cabinet was: "We're here to do whatever it takes." That was not, as some thought, a meaningless remark. It meant that in the last resort, nothing and nobody was sacred. Reagan was always ready to discard a member of the cast if he or she got in the way of the performance. The show was what mattered. He did not mind abuse. That was just the critics having their say, the reviews, as it were. What really mattered was the box office, expressed in terms of votes and public approval or disapproval of his actions. The show had to go on, and he had to carry the public with him, to rebuild American self-confidence and clout in the world so he could make it a better and safer place. He did all these things. The Reagan presidency became a hit. It ran the full eight years with growing success, and continued to resonate afterward, as an exemplar of what a presidency should be in an ideal world. But, like all careers of heroism, it was a bit of a show. Reagan's characteristics, the jokes, the benevolence, the all-American friendly hugs and squeezes and joshings were absolutely sincere. But it was acting all the same.

Margaret Thatcher worked very closely with Ronald Reagan in the 1980s. The Anglo-American alliance was never stronger than at that time, or the "special relationship" closer. They saw eye to eye on all the big international issues, and each tried to run the same kind of government at home. Thatcher, like Reagan, believed in three or four big things very strongly and both backed up their beliefs with unlimited willpower. Each proved, over and over again, that will is all-important in running a country and, in combination with a few central principles which are just, is a sure recipe for success.

Yet they were very different people, and made quite different heroes. To begin with Reagan was a very masculine man, and Thatcher a very feminine woman. She was so resolute, and occasionally so fierce, that commentators who did not know her tried to invest her with masculine characteristics. Nothing could be further from the truth. Thatcher was not only a woman to her innermost core, she loved it. I have seldom met a woman who enjoyed being a woman more. She invariably took full advantage of any feminine privilege going, from tears to tantrums, while at the same time grabbing any rights denied to women which were hers as prime minister. She loved men. Given a choice, she would always prefer being with a man to being with a woman. This manifested itself in small things, like receiving guests in a reception line. She would shake a man's hand properly. If the next in line was a woman, she would use the outstretched hand as a lever to swing the woman out of the way, so she could get to the next man. She would be civil, as a rule, to a woman who was important in her own right. But wives got short shrift.

Shortly after she became leader of the Tory party, she was a guest at a big dinner given by Woodrow Wyatt—thirty people perhaps. Her husband was not there. When the time came for the ladies to retire from the dining room, as they did in London in those days, Wyatt, a brash and shameless man, said in a loud voice: "Margaret, as you have just become, entirely on your own merits and through your own efforts, head of one of the great parties of the state, I think you should be allowed to remain with the men, at any rate for a time." One or two feminists stormed out at this juncture. Most women, even then, would have taken Wyatt's provocative remarks as a challenge to take him down a peg. But Thatcher was delighted. She moved to the end of the table where the men were congregating and called for a whiskey and water (her invariable after-dinner drink). Moreover, she did not stay "for a time." She stayed over an hour, until she felt she wanted to go home (to work, of course), and then went straight from the dining room to her waiting limo. By the time Wyatt called out, "Shall we join the ladies," she was gone.

Thatcher hated feminism. But she was not above playing the feminist card when it suited her. Equally, she hated the realm of humor since it was tricky ground for her, or she thought it was. This was another way in which she differed totally from Reagan. Her jokes were few and they tended to be somber, with a touch of bitterness. The idea of ruling through jokes, as Reagan did, was incomprehensible to her. Yet all generalizations about Thatcher tend to be faulty. She, like Reagan, was sui generis, and her behavior was never entirely predictable. In fact I once heard her make a feminist joke with stunning success. It was the fiftieth-anniversary dinner of Britain's leading economic think tank, and five hundred people, all men, had sat through a many-course, slowly served dinner, and were now listening to a series of speeches. Thatcher, prime minister, busy and overworked, had to wait until nearly midnight before being called to speak. She was livid, and I did not blame her. So was I, after two hours of all-male self-congratulatory prosing. She began:

"As the tenth speaker and the only woman, I have this to say: the cock may crow but it's the hen who lays the egg." I laughed long and loud, then suddenly realized I was the only one who thought it funny. The rest of the room sat in stunned silence.

Thatcher was a woman first. I don't recall ever seeing her wear trousers. She took more trouble over her hair than any other woman I have met. It was gold, very fine, very soft. When washed, combed and set it was magnificent. But it could come crashing down in ruin. She took immense care that this did not happen. The year after she became party leader, she and I were both invited to address the Institute of Directors at its immense annual meeting in London's Albert Hall. The dressing rooms there are below the platform, and when we joined up to make our appearance we found we had to ascend a staircase down which the air rushed as if through a wind tunnel. "My god!" said Mrs. Thatcher, "My hair has had it." "No!" said I. "Follow me. I will spread myself and you must crouch down immediately behind me." So we slowly advanced up the windy steps, like a two-headed creature, I expansive, she crouching down. On emerging, the wind-tunnel effect ceased, and she stood up, a dazzling figure, her hair perfect. "My savior!" she said to me.

She had beautiful skin, of peaches and cream, and she made herself up, lightly but effectively, with extraordinary skill, and speed. Her clothes were always fresh and immaculate, with a lot of white and royal blue. Her blouses were impeccable. Her seams were always straight, her skirts neat and perfectly in place. Whatever the time of day, however heavy the workload, in all weathers and climates, there was never the smallest sign of untidiness or disarray. She was a bandbox woman. As prime minister she was always well attended and surrounded by clever and helpful attendants, such as Carla Powell, wife of her chief foreign affairs aide, Charles Powell, and a hairdresser came to number 10 every morning. But she had always been perfectly turned out, even at Oxford, when she was Margaret Roberts of St. Hugh's College. Most girl undergraduates were scruffy or mousy, in hand-knitted woollies. But she was already perfectly groomed on all occasions when I saw her. It was a part of her nature. She was neat, clean, orderly, never flurried or breathless, in command of herself and her personal space, a prosaic poem of organized womanhood, ready for duty at all times.

Thatcher's sex was a huge disadvantage to her in her political career. The House of Commons has always been a deeply male place. It evolved out of a medieval building called St. Stephen's Chapel, and there is even something monkish about it, certainly something that seems to say, "Women are not welcome here." Though women eventually got admittance (in the 1920s), for most of Thatcher's career there were never more than thirty or forty of them in a body of over six hundred, and they always seemed anomalous in her day. The Commons is a rude and ill-bred place, with a continuous hubbub only occasionally silenced by the speaker. For a woman to make herself heard above this susurrus of sound is always difficult. If she fails to raise her voice, she is treated as inaudible and simply not listened to. If she raises her voice, the result is displeasing and she is soon described as "strident." Thatcher had to cope with this word "strident" all her career. Her voice, indeed, was the least satisfactory thing about her. It was not the voice of a heroine and could easily become the voice of a harridan. When she was entirely relaxed, her voice was perfectly acceptable. But she was rarely relaxed. Early in her public life, she developed a habit, presumably to make herself heard in the Commons, of placing heavy stress on too many words and syllables in her discourse. This had disastrous consequences, and made it impossible for her even to figure as an orator. Worse, she carried it into ordinary conversational intercourse and that made a long exchange with her (for me at least) almost unendurable. This heavy stressing went with a certain lecturing tone, which added to the agony of listening to her, particularly if her material was familiar, or even threadbare. The contrast with Reagan's voice and tone, always pleasing and often enchanting in its deft conveyance of humor, irony and persiflage, could not have been greater. Once or twice, when helping Thatcher with a speech, I tried to give her an elocution lesson, especially in getting the stresses right and in producing more varied and lively speech rhythms. I must say, she was very anxious and willing to learn, as she was in everything, if she thought her instructor knew his business. She was essentially modest (like Reagan) and tremendously keen to make herself more proficient in doing things. Nor did she have to be told anything twice, ever. She had begun her adolescence as a keen, superindustrious scholarship girl, and all her life she was an apt pupil. People said she did not listen. But that happened only when she had a low opinion of the talker. If she thought what was being said worth hearing, she not only listened hard but, on occasion, whipped out a little book from her handbag and took notes. No one was ever keener on acquiring knowledge, and correcting her faults and deficiencies. But changing a person's way of speaking is a long and difficult business, and we never had enough time together to make much progress.

Thatcher's pile-driver voice was one reason I never wanted to spend much time with her despite the fact that I was desperately anxious for her policies to succeed and did all I could in my writings to help her. She also had an irritating habit of feeding you back your own ideas. For instance, I told her once: "There are only three things a government *must* do because no one else can: external defense, internal order and running an honest currency. There are thousands of other things a government *can* do, and often does. But remember: the more optional things it does, the more likely it will neglect the three essentials." Thatcher liked this so much she wrote it down in her handbag notebook. But, lo and behold, some time afterward, when we were discussing roads, she came out with: "Remember, Paul, there are only *three things* a government *must* do . . ." and the whole rigmarole was repeated. In many ways Thatcher was a compendium of annoying habits, whereas Reagan was a compendium of pleasing ones. On the other hand, Thatcher was a good, kind and gentle creature, wonderfully considerate to her staff, always thinking of other people and doing things for them, unasked, and never cross if she got no thanks. When I told her that the Irish prime minister, whose wife was paralyzed but would not allow a live-in servant, was having to do his own laundry, she said, impulsively: "Oh, but I could go over from time to time and do it for him." Then, wistfully, "But it wouldn't do, would it?" When she went to China, and my old friend George Gale, who was covering the visit, got helplessly drunk on some sinister Chinese liquor, Thatcher put him to bed in his hotel room. Afterward she joked to me about this, and eventually I told George, who was wholly unaware of her benevolence. He thought for a few moments and said: "So *that's* why I found my clothes so neatly folded the next morning. I thought the Chinks had done it."

It is important to stress that Thatcher had a warm, kindly inner core—she was, quite simply, bighearted—because it was a side of her most people never knew existed. The great majority of heads of government, in my experience, are hardened egoists, corrupted by exercising power even if not already corrupted by getting there. The few exceptions, like Harold Wilson or Willy Brandt, tend to be weak men. To combine generosity and unselfishness with tremendous will is almost unknown, but Thatcher did, and this gave an extra dimension to her heroism. But the public, and indeed most politicians, saw only the hard, harsh, outward glitter of her battle armor. She was divisive. You either admired her enormously or detested her. This was quite unlike Reagan, a supreme egoist at the core, who inspired (an often reluctant) affection with every gesture and inflection of his voice, every aside, and who was an extraordinarily difficult man to dislike, as his opponents found. As for enemies, he never had one for long. But Thatcher tended to turn opponents into enemies, and critics into dedicated pursuers. I have an uneasy feeling, too, that she did not much mind being hated, especially by those she classified as "them," took it as part of the business of governing, and almost as proof she was on the right lines.

This propensity to arouse dislike was a serious liability, gave her hold on power always a precarious element and explains why a small group of bitter men within her party were able to dislodge her comparatively easily. On the other hand, to hold power, continuously, for nearly a dozen years, is extremely rare in British politics, and to use that tenure to accomplish fundamental changes, as she did, is rarer still. She was a born ruler, and it was a word she did not feel too shy to use. She once said to me: "I love ruling." She was also a lucky politician. The importance of luck, like the importance of humor, in politics, has been insufficiently investigated by those who write on the subject. But it is crucial. Thatcher was lucky. First, she had excellent health throughout her period of office. Second, she had been unusually lucky in getting a safe seat in North London. This saved a lot of time and energy. Third, her career in the Commons and her first steps on the ministerial ladder were marked by unusual strokes of good fortune. She got into the cabinet without making any enemies and without acquiring any awkward ideological baggage. Her capturing the leadership was a pure stroke of luck. The head of her section of the party when Ted Heath's hold on power began to slip was Keith Joseph, a gifted man whom Thatcher hero-worshiped, but who was nervous and eccentric and (I am sorry to say because I liked him too) a coward. When the time came to stand against Heath for the leadership, Joseph had a sudden spasm of funk, and pulled out. Thatcher found herself morally obliged to stand in his place, as an emergency candidate, only to discover that Heath was even more unpopular than was thought. Thus she won easily, without acquiring any reputation for pushy ambition or giving any hostage to fortune.

Her luck continued. Throughout the 1970s, public hatred of the trades unions had been growing in Britain as they ruthlessly but stupidly abused their legal powers and immunities. They destroyed the Wilson government in 1970, the Heath government in 1974 and then, in a final act of folly, they effectively destroyed the Callaghan government in the winter of 1978–1979. That precipitated Thatcher into office, in May, with a national mandate to cut the unions down to size. This she proceeded to do, with all deliberate speed and with decisive will—again, assisted by the unions themselves, who proceeded to fight two pitched battles against her, over coal and printing, battles they were bound to lose and with public opinion massed against them. I say "bound to lose" only with Thatcher as their opponent. With any other prime minister, it was by no means certain they would have lost. As it was, they enabled her to assume a heroic posture and rid the country of an unpopular scourge, once and for all. And, to cap her good fortune, a pack of idiot dictator-generals in Argentina seized the Falkland Islands and gave Thatcher the opportunity to fight a dramatic and popular war of liberation. Here again, the outcome was by no means sure. But as with the union wars, the episode gave Thatcher the opportunity to display the courage, willpower and resolution against men it was easy to hate. Thatcher was lucky in the stupidity and loathsomeness of her public enemies.

These hard-fought but straightforward and popular victories gave Thatcher the impetus to proceed in her real purpose of reducing the size of the public sector in Britain and so creating space for the entrepreneurial economy in which she strongly believed. Like Reagan, she had three or four strongly held core beliefs. But she was, unlike him, highly educated, intellectually disciplined, clear thinking and methodical. She had degrees in chemistry and law—unusually. She was widely read. She had taken in the teachings of political economists like "Fritz" Hayek, Karl Popper and Milton Friedman. She was not an intellectual. She certainly did not think that ideas were more important than people. She was not really an ideologue, more an empiricist. Her program of "privatization" evolved naturally from the mess she inherited. But, as it evolved, she saw its ideological attractions and put it in a context of theory, as it were. Thus it appealed to thinkers of the center and right all over the world, and was copied in over fifty countries. "Thatcherism" became the most popular English intellectual export since Keynesianism, which it effectively replaced.

The success of Thatcherism as a world force was also a success for capitalism itself. It gave supporters of the market system a huge burst of self-confidence, after a decade of doubt, and it was this, perhaps, as much as any other factor, which undermined correspondingly the self-confidence of the Soviet oligarchies. It was the consciousness of collectivist failure in the face of a resurgent and eager capitalism which led Gorbachev into his self-destructive attempt to "reform" Communism. This, then, was Thatcher's principal contribution to the overthrow of the Evil Empire, which could not, I think, have been accomplished without Gorbachev's nervous tinkerings. But of course at all stages Thatcher was busy stiffening the will of the Americans to be resolute over strategic weapons and to resist any nonsense from the Soviets, or anyone else. It was of immense value to Reagan to have in London a fervent supporter and ally who could always be counted upon to give him moral support, sensible advice and real friendship. It made a lot of difference to his own self-confidence. And of course the help was reciprocated. Reagan was invaluable to Thatcher both psychologically and materially during the tricky Falklands recovery operation. Thatcher's stiffening of American backbone continued after Reagan's departure. It was she who insisted that Saddam Hussein's invasion of Kuwait had to be reversed. And had she still been in power when Operation Desert Storm took place, she would never have allowed Saddam Hussein to be let off the hook but would have obliged George Bush Sr. to replace his regime. Thus the second Iraq war would have been avoided. But such hypothetical speculations are useless, and painful.

Thatcher was the only British leader since Churchill to have a perceptible influence on world events, both directly and through her high standing in Washington. Her claim to heroic status is unquestionable. And she liked heroes herself. She tended to see the world in black-and-white terms, and to label the current cast "goodies" and "baddies." She picked up the old White's Club bar term "wets" and applied it fiercely to those who lacked the guts to follow her in her adventures. She loved those who wholeheartedly committed themselves to a resolute course of righteous action. That was why she loved Pope John Paul II, the third member of the blessed trinity of heroes who destroyed the Communist monolith. I shall always regret the mischance which prevented Thatcher and myself going to see the pope together. It was after she left number 10 but while John Paul was still at the height of his energy. Thatcher and I were in Rome for the opening of a hotel restored by our friend Olga Polizzi. An interview with the pope had been arranged, but he postponed it at the last moment in order to finish his encyclical on sex, which he regarded as one of the most important things he had ever written. Instead he invited us to a private showing of the Sistine Chapel. So Thatcher and I went around it together, and we promptly got into a furious argument about whether Michelangelo was a homosexual or not. Thatcher enjoyed arguments, and so do I. But the Italian officials and monsignori could not distinguish between a noisy argument and a quarrel, and cowered in fear and horror.

John Paul II may have been the most important of the blessed trinity because he understood the Soviet empire on the ground in Poland. By giving his moral and, to some extent, physical support to the trades union movement of Lech Walesa, and by making himself the active spiritual leader of the united Polish people, he undermined the empire fatally. Once his ghostly leadership on the actual soil of Poland was firmly established, there was never any possibility of Soviet imperial rule reestablishing itself without a bloodbath of a kind not even Brezhnev would have been prepared to face, and all his successors funked totally. In many ways it was the most impressive display of papal political power since the time of Innocent III in the early thirteenth century, and gave the true answer to Stalin's brutal (and foolish) question: "How many divisions has the pope?"

However, I don't intend to dwell at length on John Paul because he inhabits the borderland between the heroic and the saintly, and this is a book about heroes, not saints. I first glimpsed him as he was about to enter the conclave which elected him, and he was pointed out to me as *papabile.* He seemed then a formidable physical presence rather than an ethereal one, a man of rugged, sturdy limbs and considerable muscular power, who had kept himself in trim by skiing and mountaineering. But, invested with papal authority and the charisma which goes with this unique office, he rapidly acquired an extra dimension of personality and his appearances became epiphanies. Thus I saw him at the height of his spiritual grandeur when he came to Canterbury to pay tribute at the shrine of St. Thomas à Becket, a man he resembled in more than one respect. He appeared to dwarf the large cluster of dignitaries, lay and ecclesiastical, who surrounded him, as though he came from a different species. I gazed down at him, from my superb seat high in the vault of the sublime cathedral, with wonder and awe. No other man I have seen ever emitted such radiance.

As a man, as an ecclesiastical statesman, John Paul possessed three qualities which can truthfully be described as heroic. First, from the outset of his pontificate he showed a willingness to take on a monumental task. During that disastrous decade, the 1960s, when so many institutions began a course of self-destruction, the Vatican also succumbed. Pope John XXIII, with the best intentions, and out of the sweetness and innocence of his nature, summoned the Second Vatican Council, and bid it "open windows" and renew the Church. Then he died before he could supervise its entire deliberations and rein in its consequences. Under his successors, especially the saturnine and accident-prone Paul VI, the post-councilian juggernaut of change got completely out of control. It was John Paul's historic and divine mission to get the Church back under control, an enormous and frightening task, which he set about with enviable courage, fortitude and determination. I have already described what he did in an earlier book, and here all I would say is that by persistence over the long haul and by the obstinacy of a St. Jerome in the desert, he did the job, and left the church not only stable and at rest but ready to appoint an admirable successor.

John Paul's second heroic quality was the purity of his behavior. By this I mean the way in which he carried through a singularly difficult transformation of a complex institution, involving a good deal of political maneuvering and skill, and delicate (occasionally forceful) handling of senior individuals, without any sacrifice of principle or descent into unworthy tactics. I never heard him accused of behavior which would not have passed muster at a synod of moral theologians. Bearing in mind the skull-duggery traditionally associated with the internal affairs of the Vatican, this is a remarkable record. It is not that John Paul had no critics. On the contrary. In some ways he was a divisive figure, like Thatcher. Many liberal Catholics hated him and opposed him bitterly in the journals to which they had access, or controlled. But it was his aims they objected to, not the methods by which he secured them, which were always straightforward and honest.

Third was John Paul's singleness of purpose. He was heroic in his simplicity, which went hand in hand with great intellectual sophistication, profound knowledge and flexibility of argument. He never allowed himself to be deflected from his restorative program, which he pursued steadily and tenaciously throughout his long pontificate. He had, it seemed to me, a strong sense of priorities, an unfailing ability to separate the essential from the peripheral, and to keep to the point, obliging others to do likewise. His intellect was burly, gripped hard and never relaxed until the job was done.

Finally, his death was heroic. It was the *bona mors* raised to a monumental level. He showed us all, including future popes yet born, how to die. He worked, ruled, read and, above all, prayed right to the end. He never gave in. There was no surrender except, finally, to the imperious summons of his maker. He remained the vicar of Christ so long as there was breath in his body. Those last weeks of a dying pope, followed by a funeral which might have been managed by Michelangelo, will remain in the memory as a heroic episode in itself. We expect popes to be saints, not heroes. But John Paul was both.

EPILOGUE

ANY FUTURE FOR HEROISM?

Heroes have not always been appreciated. Indeed, in the troubled times which followed the end of the Napoleonic Wars, caused by lower wages, unemployment and high food prices, the term became abusive. Whenever the Duke of Wellington made his appearance, a certain kind of London mob (there were many different kinds) would shout: "No heroes! We want no heroes!" For the self-conscious proletariat, the "Man on Horseback" was a political enemy. They threw stones through the windows of Apsley House, the duke's London residence. He had the windows boarded up but refused to replace the glass, as a reminder to people of how volatile was popularity and how fickle the crowd, applauding him as a hero one moment, detesting him the next.

The truth was that the duke, being not so much a cynic as a realist, a man who had no illusions about human nature, disapproved of the hero cult. He had seen what it led to in France, where the egregious hero of the age, Bonaparte, had been able, in consequence, to send millions—friends, enemies and civilians—to their deaths. It had taken all his efforts, and much blood, to stamp out this particular outbreak of hero worship. What the duke was not to know, but might have foreseen and deplored, was that the return of Bonaparte's body to France (which he opposed) and its sanctification in the Invalides, the world's most vainglorious tomb, led to his posthumous cult of grandeur and aggression. Hence the absurd dictatorship of his nephew, Napoleon III, a disaster for France, ending in the disgrace of Sedan and the bloodbath of the Paris Commune, and indeed for Europe too, for it made possible the military paramountcy of Prussia, and so pointed directly to the horrors of the First World War. Nor did this evil heroic streak end in 1918. On the contrary, the age of the dictators was a consequential echo of the age of Napoleon. For these self-made autocrats, beginning with Atatürk in the debris of the Ottoman Empire, and Mussolini in Italy, and continuing through Stalin in Russia, Hitler in Germany and Mao Tse-tung in China, all to some degree were inspired by, and modeled themselves on, Bonaparte, the prototype and exemplar of all modern tyrants. Many lesser monsters followed in their wake: Sukarno in Indonesia, Ho Chi Minh and Pol Pot in Indochina, Nasser and Saddam in the Middle East, Peron and Castro in Latin America, Gadhafi, Bokassa and Idi Amin in Africa—all of whom strutted about for a time in grotesque parodies of the masterful Corsican, engaging in every form of beastliness, from genocide to cannibalism.

Wellington, who had studied and taken to heart Roman history, and the use of emperors placed on the throne by their troops, as well as the disastrous example of Bonaparte, believed that the essence of a constitutional state was the absolute submission of the military power to duly constituted civil authority. He strongly approved of Washington's conduct in the delicate transmission from war to peace, and the creation of the American constitutional government, just as, equally, he deplored the example of Simón Bolivar in South America, which set that unhappy continent on its tragic path to periodic coups d'état and military rule. Thanks to Washington's wisdom and forbearance, the United States has never been in real danger of taking the Bonapartist path, always—as in the case of General MacArthur and President Truman half a century ago—the military man submitting to the elected chief magistrate, with the full approval of the nation.

Wellington's dislike of professional heroes had another aspect to it. He disliked showiness, or "side," as they called it at Eton, and, still more, boasting. A particular object of his contempt was Admiral Sir Sydney Smith, the so-called or self-proclaimed hero of Acre, who had prevented the Turks from taking this valuable port and had dined out on it ever since. His windiness on the subject led to his attracting the nickname of "Long Acre," after a district of Central London. The admiralty secretary, Croker, warned Wellington about Smith in 1814: "An old naval friend told me: 'Beware of heroes—the more you come to know them, the less you will think of them.'" Smith was a loose cannon during the Congress of Vienna, and for a time attached himself to Wellington. But the duke soon got rid of him: "Talks too much."

The scale on which military heroes have been rewarded has declined. In England, the nation gave the Duke of Marlborough, victor in four major battles against Louis XIV, Blenheim Palace, still the most princely house in the country. Wellington declined anything on that scale, but was provided with the fine estate of Stratfield Saye in Hampshire. Even Field Marshal Haig, dubious hero of the western front in the First World War, was given £100,000 by Parliament. By contrast, Field Marshal Sir Alan Brooke, chief of staff under Churchill for most of the Second World War, got a "gratuity" of £346, and had to sell his house and his beloved collection of bird books, just to make ends meet when peace came. None of the British heroes of that war were financially rewarded, except by selling their memoirs. America has never gone in for such prizes. But it sometimes bestows, democratically, high office on its heroes—Grant and Eisenhower being examples.

The twentieth century, and still more the twenty-first, has followed in the footsteps of the nineteenth and broadened the recruiting ground for heroes. Men like Humphry Davy, inventor of the miner's safety lamp, or Thomas Edison, the world's inventor in chief, came on to the stage of public adulation followed by explorers like Stanley and Livingstone, and brave women in the shape of Florence Nightingale and Grace Darling. There were hero praisers too: in Britain Carlyle stuck to the traditional mode of the captains, like Cromwell and Frederick the Great, but in America Emerson prepared the way for the cult of the entrepreneur. In due course the public applauded the outstanding steelmaker Andrew Carnegie, and the oilman John D. Rockefeller.

This new kind of hero was controversial, and it is a fact that, throughout history, one person's hero has been another's villain, not only in his own day but later. I recall an elderly don of my old Oxford college coming into the Senior Common Room one evening and announcing to his colleagues: "You know, I am really coming to detest that fellow Julius Caesar"—as though he were about to stride through the door, laurel on brow, sword in one hand and *Gallic War* in the other. The heroes of America's emergence as the world's largest industrial power were clearly genuine in one sense, since Carnegie's cheap, high-quality steel benefited everybody; Rockefeller's slashing of the price of kerosene by 90 percent was a godsend to the housewife; and Ford's cheap, reliable Model T ended the isolation of the farmer. But to others, such men were "robber barons," or, in the words of President Theodore Roosevelt, "malefactors of great wealth." The rich sought redemption through philanthropy. Carnegie declared that amassing wealth was morally allowable, if done honestly, "but he who dies wealthy dies disgraced." He gave away his hundreds of millions in an endless series of benefactions whose catalog alone fills a fair-size book. The Ford and Rockefeller Foundations followed in due course, and the series continues today with the colossal benefactions of Bill Gates and Warren Buffett. Now people feel that giving away money in vast quantities is not heroic, if your wealth is colossal. I am not so sure. In my experience, a majority of rich people are stingy. Of course the real hero is someone like Dr. Johnson, who always gave away a large chunk of his income, and would fill his pockets with coins before going out, entirely for the purpose of donating small sums to beggars. "What is the point of it?" he was asked. "To enable them to *beg on*," he replied, a truly heroic remark. The generous poor man, like Charles Lamb, is a particular hero of mine. But I must admit I enjoy the company of the generous rich, like my old friend Jimmy Goldsmith, and don't feel it is right to inquire too closely as to how they came by their wealth.

People must agree to differ about heroes. I admire Chile and its people greatly, and became concerned when my friend Salvador Allende became its president and opened the country to hordes of armed radicals from all over the world. The result was the world's highest inflation, universal violence and the threat of civil war. So I applauded the takeover by General Pinochet, on the orders of Parliament, and still more his success in reviving the economy and making it the soundest in Latin America. But by preventing the transformation of Chile into a Communist satellite, the general earned the furious hatred of the Soviet Union, whose propaganda machine successfully demonized him among the chattering classes all over the world. It was the last triumph of the KGB before it vanished into history's dustbin. But Pinochet remains a hero to me because I know the facts.

My other heroes tend to be people who successfully accomplish things I would not dare even to contemplate. I could not possibly sail single-handedly round the world, even if I had the skill, like a pretty and fragile woman of my acquaintance, Clare Francis. My admiration for her is without qualification, particularly since she is modest about her achievement. The man who runs a fruit stall round the corner from my house has swum the English Channel several times for charity. He is a true hero for me. I admire heroines of the oriental slums like Mother Teresa, who was a realist as well as an idealist (as are most true saints). The vicious attacks launched on her by Western playboy intellectuals filled me with horrified fury. I always have a soft spot for those who speak out against the conventional wisdom and who are not afraid to speak the truth even if it puts them in a minority of one. And in this case there is some common feeling, for during most of my life I have been outspoken myself on these lines, and have suffered accordingly. I think we appreciate heroism most if we have a tiny spark of it ourselves, which might be fanned into a flame if the wind of opportunity arose.

So how do we recognize the heroes and heroines of today? I would distinguish four principal marks. First, by absolute independence of mind, which springs from the ability to think everything through for yourself, and to treat whatever is the current consensus on any issue with skepticism. Second, having made up your mind independently, to act—resolutely and consistently. Third, to ignore or reject everything the media throws at you, provided you remain convinced you are doing right. Finally, to act with personal courage at all times, regardless of the consequences to yourself. All history teaches, and certainly all my personal experience confirms, that there is no substitute for courage. It is the noblest and best of all qualities, and the one indispensable element in heroism in all its different manifestations.

NOTES ON FURTHER READING

Among works on women heroes of the Old Testament, Deborah is discussed in Micke Bal, *Murder and Difference: Gender, Genre and Scholarship in Sisera's Death* (Bloomington, 1988), and in L. L. Bronner, "Valorized or Vilified? The Women of Judges in Midrashic Sources," in *A Feminist Companion to the Book of Judges*, ed. Athalya Brenner (Sheffield, 1993). For Samson, see the writings of J. Cheryl Exum, "Aspects of Symmetry and Balance in the Samson Saga," in *Journal for the Study of the Old Testament* 19 (1981) and "The Theological Dimensions of the Samson Saga," in *Vetus Testamentum* 33 (1983). Good annotated editions of Judges are R. G. Boling, *Judges, The Anchor Bible* (New York, 1975) and J. A. Soggin, *Judges: The Old Testament Library* (Philadelphia, 1981). The Anchor Bible has a good volume on Judith by C. A. Moore, and J. C. Vanderkam has edited a set of essays on her: *No One Spoke Ill of Her* (Atlanta, 1992).

For Alexander there is a bibliography in volume 6 of the *Cambridge Ancient History*, pp. 529ff. The best modern life is by Lane Poole, but I also recommend G. T. Griffith (ed.), *Alexander the Great: The Main Problems* (London, 1965). For Caesar see J. P. V. D. Balsdon, *Julius Caesar* (London, 1967), but Mommsen, *History of Rome* (Eng. trans. 1895), vols. 4–5, is still valuable. I also recommend the English translation by David McLintock of Christian Meier's *Julius Caesar* (London, 2004).

For Boudica, the best modern biography is Richard Hingley and Christina Unwin, *Boudica: Iron Age Warrior Queen* (London, 2005), but M. J. Trow, *Boudica: The Warrior Queen* (Stroud, 2003) is also useful. For Joan of Arc, the most valuable source is the record of the *Procès de condamnation et de réhabilitation*, ed. Jules Quichevat, 5 vols. (Paris, 1861), and a modern edition by P. Duparc, 5 vols. (Paris, 1977–1988). There is an excellent volume on Joan of Arc by Vita Sackville-West, first published in 1936; I have used the 2004 edition. Modern biographies include those by the French scholar R. Pernond (trans. 1961) and, in English, by Lucie Smith (1976), Warner (1981) and Barstow (1986).

For the Tudor and Jacobean heroes, the best modern life of Sir Thomas More I know is by Richard Marius (New York, 1984). Lady Jane Grey's *Letters* were edited by D. Geary (Ilfracombe,1951); there is an old, but good, *Life* by R. P. B. Davey (London, 1909). N. H. Nicolas edited *The Literary Remains of Lady Jane Grey* (London, 1825). For Elizabeth I, see the many works of J. E. Neale, and my *Elizabeth: A Study in Power and Intellect* (first published London, 1974). On Ralegh, see A. L. Rowse, *Ralegh and the Throckmortons* (London, 1962) and Robert Lacey, *Sir Walter Ralegh* (London, 1973). Antonia Fraser, *Mary Queen of Scots* (London, 1991 edition) is still the best life of Mary.

For Washington there is a famous seven-volume biography by Douglas Southall Freeman (New York, 1948–1954), with a one-volume condensation by Richard Harwell (1968). The best recent biography are the two large volumes by Harrison Clark, *All Cloudless Glory: The Life of George Washington* (Washington, D.C., 1996). For brevity, see my *Washington: The Founding Father* (New York, 2005). For Nelson, a good modern life is Edgar Vincent's *Nelson: Love and Fame* (London, 2003). Also excellent is Christopher Hibbert, *Nelson: A Personal History* (London, 1995). The best life of Wellington is by Elizabeth Longford in two volumes (London, 1969–1972). I also recommend the essay by John Keegan in his book *The Mask of Command* (London, 1987) and E. A. Smith's *Wellington and the Arbuthnots* (Stroud, 1994). *Wellington's Conversations with Stanhope* is available in many editions.

Jane Welsh Carlyle is best studied through her letters. Indispensable is K. J. Fielding and D. R. Sorensen, *Jane Carlyle: Newly Selected Letters* (Aldershot, 2004); the complete edition of the letters by both Carlyles is the huge project conducted by Edinburgh and Duke University, North Carolina, of which thirty-four volumes have so far been published, carrying the story up to December 1858. A recent life of Carlyle is Simon Heffer, *Moral Desperado* (London, 1995). An indispensable book on Dickinson is Alfred Habegger: *My Wars Are Laid Away in Books: The Life of Emily Dickinson* (2001).

Of modern biographies of Lincoln the one I recommend is David Herbert Donald, *Lincoln* (New York, 1995). But I also like the two-volume *Abraham Lincoln: Speeches and Writings*, published in the Classics of America Library (New York, 1992). Valuable is Gary Wills, *Lincoln at Gettysburg* (New York, 1992). For contrast, William C. Davis, *Jefferson Davis: The Man and His Hour* (New York, 1991) is instructive. For Robert E. Lee, the best modern biography is Emory M. Thomas, *Robert E. Lee: A Biography* (New York, 1995).

For Wittgenstein, the best biography is Roy Monk, *Ludwig Wittgenstein: The Duty of Genius* (London, 1990). There are many memoirs of him by colleagues and pupils; indeed I have looked through more than a hundred books on this man. A good study is W. W. Bartley III, *Wittgenstein* (London, 1986 edition) and a racy monograph is David Edmonds and John Eidinow, *Wittgenstein's Poker* (London, 2001) about his famous row with Karl Popper. Anthony Kenny has edited *The Wittgenstein Reader* (Oxford, 1994), and there is an amusing picture book by John Heaton and Judy Groves, *Wittgenstein for Beginners* (Cambridge, 1994).

For Churchill it is necessary to master the eight-volume biography begun by his son Randolph but written very largely by Martin Gilbert. This is accompanied by innumerable companion volumes giving the full texts of letters and papers from the Churchill archives. De Gaulle's *Memoirs* are better reading and no less mystifying than any of the many biographies of him.

Mae West has the benefit of an excellent biography of her, Simon Louvish, *Mae West: It Ain't No Sin* (London, 2005), based on her extensive personal archives. There is, alas, no equivalent life of Marilyn Monroe, and I have had to rely on ephemeral screen biographies, clippings, and personal reminiscences of those who knew her.

Ronald Reagan has so far benefited from one first-class biography, Lou Cannon, *The Role of a Lifetime* (New York, 1991). There is no equivalent for Margaret Thatcher, since the official life by Charles Moore will not be published until after her death. I have relied on my own archive, including diaries, letters and memories. For Pope John Paul II see my study, *John Paul II and the Catholic Restoration* (New York, 1982). His encyclicals, covering a vast range of subjects, are the best entry into his mind and faith.

INDEX